Value Creation through Engineering Excellence

Yufeng Zhang · Mike Gregory
Editors

Value Creation through Engineering Excellence

Building Global Network Capabilities

palgrave
macmillan

Editors
Yufeng Zhang
Birmingham Business School
University of Birmingham
Edgbaston, Birmingham
UK

Mike Gregory
Department of Engineering
University of Cambridge
Cambridge
UK

ISBN 978-3-319-56335-0 ISBN 978-3-319-56336-7 (eBook)
https://doi.org/10.1007/978-3-319-56336-7

Library of Congress Control Number: 2017939886

Printed on acid-free paper

This Palgrave Macmillan imprint is published by Springer Nature
The registered company is Springer International Publishing AG
The registered company address is: Gewerbestrasse 11, 6330 Cham, Switzerland

Preface

Engineering plays a central role in economies and industries. Its influence goes far beyond the recognisable sectors such as automotive and aerospace. Engineering expertise underpins industries as diverse as food and entertainment, retail and pharmaceuticals. But the 'invisibility' of much engineering activity may lead to its significance being underestimated. At a time when new technologies and global economic structures are changing so rapidly this book sets out to explore the role and evolution of engineering in different contexts.

The focus of the book is on those engineering activities which create substantial value. It highlights those activities that are essentially innovative and lead to growth of businesses and improvements in living standards rather than those which simply maintain the status quo—important as that remains. The term High Value Engineering has been coined to characterise these activities though it is recognised that this is an evolving definition and indeed activity.

The book reflects an extended collaboration between leading scholars across Europe and China and draws upon leading research programmes while remaining firmly grounded in engineering practice.

The early chapters explore new approaches to the representation of engineering which it is hoped will provide new insights into the way modern engineering is performed. The recognition of an engineering value chain and the highly networked nature of engineering activities around the world are thought to be particularly important in recognising and managing the role of engineering within companies and indeed nations.

Subsequent chapters address the characteristics of different aspects of modern engineering from design to services. Concepts of value chains and value creation are explored in greater detail demonstrating engineering in its broader business and commercial context. The opportunities and challenges presented by rapid developments in ICT are considered as well as the implications for skills and training. Increasingly important issues surrounding sustainability are tackled recognising again the critical role of engineering in moving towards the environmentally sustainable production and use of goods and services.

The book concludes with a chapter on policymaking with respect to engineering and technology. The complex skills and expertise associated with High Value Engineering represent a precious asset not just for companies but for society more generally. It is essential that these resources are deployed sensibly and knowingly with an understanding their potential and limitations.

The authors earnestly hope that this book will provide new insights into the nature and role of modern engineering, informing industrialists and policymakers and perhaps most importantly enthusing younger scientists, technologists and engineers with excitement for the rapidly developing field of High Value Engineering.

Edgbaston, Birmingham, UK — Yufeng Zhang
Cambridge, UK — Mike Gregory

Acknowledgements

The authors would like to express their gratitude to the support of the Seventh Framework Programme of the European Union through Marie Curie Actions IRSES Europe-China High Value Engineering Network (EC-HVEN), http://cordis.europa.eu/result/rcn/191425_en.html. The authors would also like to thank Chris Collinge and Amorettya Minayora for their kind helps of proofreading the chapters.

Contents

Editors and Contributors

About the Editors

Yufeng Zhang is Senior Lecturer of Operations Management at the Business School of the University of Birmingham, UK. He had an earlier industrial career with responsibilities for project management and new business development, and continues to work very closely with industry aiming to help companies cope with challenges and opportunities in managing complex global engineering networks. He has researched and published in the areas of engineering management, network operations, international manufacturing, technology management and innovation.

Mike Gregory is Emeritus Professor of Manufacturing at the University of Cambridge Engineering Department., UK. He chairs the UK Manufacturing Professors Forum, co-chairs the Manufacturing Advisory Group of the UK Government's Department of Business Innovation and Skills (BIS), and is a member of the supervisory board of the High Value Manufacturing Catapult. Mike is a Fellow of Churchill College, Cambridge, and of the Royal Academy of Engineering, UK, where he is a trustee.

Contributors

Saeema Ahmed-Kristensen Imperial College London, London, UK

John R. Bryson University of Birmingham, Birmingham, UK

Chris Collinge Birmingham Business School, Birmingham University, Birmingham, UK

Ming Dong Shanghai Jiao Tong University, Shanghai, China

Steve Evans Institute for Manufacturing, Cambridge University, Cambridge, UK

Mike Gregory Department of Engineering, University of Cambridge, Cambridge, UK

Ting He College of Computer Science and Technology, Huaqiao University, Xiamen, China

Yonglin Li Shanghai Jiao Tong University, Shanghai, China

Nichola Lowe University of North Carolina, Chapel Hill, USA

Veronica Martinez Institute for Manufacturing, Engineering Department, Cambridge University, Cambridge, UK

Tim Minshall Institute for Manufacturing, University of Cambridge, Cambridge, UK

Letizia Mortara University of Cambridge, Cambridge, UK

Rachel A. Mulhall University of Birmingham, Birmingham, UK

Andy Neely Institute for Manufacturing, Engineering Department, Cambridge University, Cambridge, UK

Rob Phaal Institute for Manufacturing, University of Cambridge, Cambridge, UK

David Probert Institute for Manufacturing, University of Cambridge, Cambridge, UK

Padmakshi Rana Institute for Manufacturing, Cambridge University, Cambridge, UK

Yongjiang Shi Institute for Manufacturing, Engineering Department, Cambridge University, Cambridge, UK

Julianne Stern University of North Carolina, Chapel Hill, USA

Erik Søndergaard Technical University of Denmark, Lyngby, Denmark

Jing Tao School of Mechanical Engineering, Shanghai Jiao Tong University, Shanghai, China

Thomas P. Taylor Technical University of Denmark, Lyngby, Denmark

Florian Urmetzer Institute for Manufacturing, Engineering Department, Cambridge University, Cambridge, UK

Xiaobo Wu School of Management, Zhejiang University, Hangzhou, China

Xiaofei Xu School of Computer Science and Technology, Harbin Institute of Technology, Harbin, China

Miying Yang Institute for Manufacturing, Cambridge University, Cambridge, UK

Suiran Yu School of Mechanical Engineering, Shanghai Jiao Tong University, Shanghai, China

Tao Zhang Birmingham University, Birmingham, UK

Yufeng Zhang Birmingham Business School, University of Birmingham, Birmingham, UK

Lina Zhou School of Computer Science and Technology, Harbin Institute of Technology, Harbin, China

Quan Zhou School of Management, Zhejiang University, Hangzhou, China

List of Figures

Chapter 12

List of Tables

Part I

The Engineering Value Chain and the main Capability Areas

1

Introduction

Yufeng Zhang and Mike Gregory

1 High Value Engineering (HVE), the Engineering Value Chain (EVC), and Global Engineering Networks (GEN)

Engineering directs the sources of power in nature for the benefit and convenience of human being. The subject has been broadly considered as the synthesising practice of creative art and scientific technology, and specifically defined as problem solving within the constraints of technology, material, budget, and time. High value engineering (HVE) in this context refers to the application of engineering technologies, skills, and capabilities in creating high-value-added products and services that

Y. Zhang (✉)
Birmingham Business School, University of Birmingham, Birmingham, UK
e-mail: zhangys@bham.ac.uk

M. Gregory
Department of Engineering, University of Cambridge, Cambridge, UK
e-mail: mjg@eng.cam.ac.uk

Y. Zhang and M. Gregory (eds.), *Value Creation through Engineering Excellence*,
https://doi.org/10.1007/978-3-319-56336-7_1

may lead to sustainable economic growth in complex global business networks. Value creation through engineering excellence focuses on the successful transformation of novel ideas and new technologies into marketable products and services as well as the effective integration of a series of engineering activities dispersed in various industrial settings and along the engineering value chain (EVC). The EVC consists typically of five categories of activities through which engineering operations contribute to customer value and thus the overall competitiveness of a company—idea generation and selection, design and development, production and delivery, service and support, and recycling and disposal (Zhang and Gregory 2011).

Competition in the modern global economy relies on gaining superior engineering capabilities to create high-value-added products and services rather than focusing solely on the output of an engineering system. Such outputs are bound in local contexts and less effective, serving international markets. Engineering network capabilities will address that limitation because they are adaptive and can be transferred across geographic and organisational boundaries. Global engineering operations are thus gaining an increasing importance driven by the rapid growth of the emerging economies, increasing engineering capabilities (and workforce) in the developing countries, the global race for talent, and opportunities made available by the progress of information technologies (Zhang and Gregory 2013). Engineering companies are pioneering new forms of organisations which are knowledge based, technology enabled and globally networked, as a result of the nature of the knowledge they deploy, the degree of jurisdictional control they exercise, and the global client relationships they seek. From a knowledge perspective, engineering companies tend to adopt lateral team structures and reciprocal processes since they have a technical or syncretic knowledge base supported by multiple disciplines rather than a normative knowledge base. From a jurisdiction perspective, engineering professions have weaker social closure and looser geographic jurisdictional boundaries; therefore, it is relatively easy for engineering companies to form a global network structure. From a client perspective, engineering companies require a high degree of face-to-face client interaction in the production process, and thus a high degree of geographic dispersion of assets. The design and operations of global engineering networks (GEN) are expected to develop these essential capability elements by accessing and deploying dispersed resources, integrating and coordinating networked activities, and managing engineering knowledge and collective learning (Zhang et al. 2016).

Global engineering networks (GEN) are knowledge intensive, people-centric, and very often project based. As a result of these distinctive features, the primary concern of engineering operations differs from that of manufacturing or basic research in the tasks, outputs, and required knowledge. The main drivers for engineering task choice are usually external sources rather than the internal curiosity of an engineer or the scientific desire of an engineering organisation. The outputs are often one-off designs or solutions, rather than standardised manufacturing outputs, or a scientific enquiry or theory purely to improve our understanding of the world or to fulfil the discovery desire of a researcher. The required knowledge, especially engineering know-how, is often intangible and embedded in different parts of an organisation or a group of organisations. Such intangible, practical, unpredictable, and embedded characteristics should be properly addressed in building high value engineering capabilities in an international context (Zhang et al. 2014).

2 An Overview of the Book

This timely book provides a holistic view of this frontier knowledge area for building high value engineering capabilities in global network operations. It updates the traditional disciplines of engineering and operations management by addressing challenges and opportunities in building global network capabilities. It also addresses a critical problem of the relevant subject areas which are either technically driven or incapable of dealing with the increasing complexity and dynamics in global engineering networks. The readership includes researchers, practitioners, policymakers, and students working in a wide range of high value engineering areas, especially engineering management, operations management, service operations, supply chain management, technology management, innovation management, engineering design and

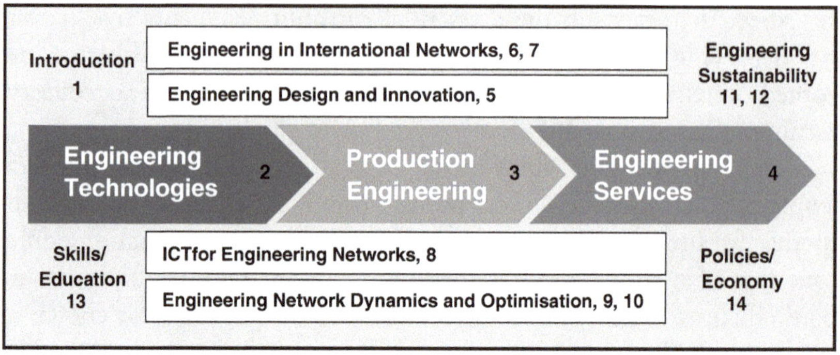

Fig. 1 An overview of the book

knowledge management, international manufacturing, industrial sustainability, industrial policy, and regional economy.

The book contains a set of comprehensively developed chapters organised into three parts (Fig. 1: An overview of the book). The first part outlines the engineering value chain and introduces the main capability areas of high value engineering. This section reveals the changing global landscape of engineering, and provides a systematic view of engineering capabilities which are critical to global competition in the current business environment and in the future. To set a scene for discussions in the following parts, key capability areas are introduced around engineering technologies—to explore the linkage between basic research and emerging technologies to high value engineering; production engineering—to explain the implications of advanced manufacturing methods to high value engineering; and engineering services—to illustrate the value creation and transformation processes among engineering services. The part has four chapters.

Part I: The Engineering Value Chain and the Main Capability Areas

- Chapter 1: Introduction
- Chapter 2: Engineering and Technology Management
- Chapter 3: International Manufacturing and Engineering
- Chapter 4: Engineering Services—Unpacking Value Exchange

The second part of this book includes latest methodologies and technologies to support innovation and optimisation in complex global engineering networks. Such developments require the effective integration of engineering capabilities along the whole engineering value chain. Focusing areas of discussion include engineering design, global product development, innovation in international networks, information and communication technologies (ICT), as well as modelling techniques for engineering network coordination, simulation, and optimisation.

Part II: Engineering Network Innovation and Optimisation

- Chapter 5: Engineering Design and Innovation in a Global Context
- Chapter 6: Engineering in International Business Networks
- Chapter 7: Engineering Value Chain Simulation and Innovation
- Chapter 8: ICT for High Value Engineering Networks
- Chapter 9: Engineering Value Chain Modelling and Optimisation
- Chapter 10: Engineering Value Chain Coordination and Optimisation

The third part of this book looks into future trends and discusses implications for developing high value engineering capabilities from the broader perspectives of industrial sustainability, skills, education, economy, and industrial policies in an international context.

Part III: Future Trends and Implications

- Chapter 11: Engineering for Sustainable Value
- Chapter 12: Product Life Cycle Design for Sustainable Value Creation
- Chapter 13: Engineering and the Skills Crisis in the UK and USA
- Chapter 14: An Industrial Policy Framework for High Value Engineering

All together, our contributors have completed this book to place value creation through engineering excellence on a solid ground of scientific theories and enabling techniques. This will help engineers, scientists, managers, and students to gain an overall understanding of this cross-disciplinary knowledge area of an increasing importance in the modern economy. Such an overall understanding of high value engineering will enable companies of different sizes to benefit from global engineering networks as well as informing policymakers to develop effective industrial policies for enhancing the competitiveness of engineering sectors at regional and national levels. It has also been expected that our collective efforts will lead to a systematic research agenda for further advancement in this knowledge area through studying key issues in building global engineering network capabilities.

References

Zhang, Y., & Gregory, M. (2011). Managing global network operations along the engineering value chain. *International Journal of Operations & Production Management, 31*(7), 736–764.

Zhang, Y., & Gregory, M. (2013). Towards a strategic view of engineering operations. *Proceedings of the Institution of Mechanical Engineers, Part B: Journal of Engineering Manufacture, 227*(6), 767–780.

Zhang, Y., Gregory, M., & Shi, Y. (2014). Managing Global Engineering Networks (GEN) Part I: Theoretical foundations and the unique nature of engineering. *Proceedings of the Institution of Mechanical Engineers, Part B: Journal of Engineering Manufacture, 228*(2), 163–180.

Zhang, Y., Gregory, M., & Neely, A. (2016). Global engineering services: Shedding light on network capabilities. *Journal of Operations Management, 42* (3), 80–94.

Authors' Biography

Yufeng Zhang is Senior Lecturer of Operations Management at the Business School of the University of Birmingham, UK. He had an earlier industrial career with responsibilities for project management and new business development, and continues to work very closely with industry aiming to help companies cope with challenges and opportunities in managing complex global engineering networks. He has researched and published in the areas of engineering management, network operations, international manufacturing, technology management and innovation.

Mike Gregory is Emeritus Professor of Manufacturing at the University of Cambridge Engineering Department., UK. He chairs the UK Manufacturing Professors Forum, co-chairs the Manufacturing Advisory Group of the UK Government's Government's Department of Business Innovation and Skills (BIS), and is a member of the supervisory board of the High Value Manufacturing Catapult. Mike is a Fellow of Churchill College, Cambridge, and of the Royal Academy of Engineering, UK, where he is a trustee.

2

Engineering and Technology Management

Yuan Zhou, Rob Phaal, Tim Minshall and David Probert

1 Getting Value from Engineering and Technology

The impact of engineering and technology is visible in the many artefacts, infrastructure and services that form the fabric of modern life. The industrial revolution in the eighteenth century enabled new manufacturing processes that have transformed the world, spurred on by continual technological developments, both incremental and transformational. Scientific breakthroughs have enabled new technologies that disrupt

Y. Zhou (✉)
School of Public Policy and Management, Tsinghua University,
Beijing, China
e-mail: zhou_yuan@mail.tsinghua.edu.cn

R. Phaal · T. Minshall · D. Probert
Institute for Manufacturing, University of Cambridge, Cambridge, UK
e-mail: rp108@cam.ac.uk

D. Probert
e-mail: drp1001@cam.ac.uk

Y. Zhang and M. Gregory (eds.), *Value Creation through Engineering Excellence*,
https://doi.org/10.1007/978-3-319-56336-7_2

existing industries (Bower and Christensen 1995) and enable the creation of new ones. Technological disruption will bring great benefits but also challenges, such as the impact of computerisation on skills and employment (Frey and Osborne 2013).

Scientists, engineers and other technologists have a crucial role to play in the development and deployment of technology for the economic benefit of society, and also to address challenges facing humanity, such as climate change and resource scarcity. In this context, the management of engineering and technology becomes increasingly important. Technological investment and effort needs to be aligned with organisational and wider social needs and aspirations throughout the life cycle from design, through to production and the creation of valuable services, as depicted in Fig. 1.

The word "engineer" is derived from the Latin ingeniare, meaning "to produce" (Mitcham 1978). This original form of interpretation highlights the many activities and roles that engineers undertake, deploying scientific and craft knowledge to create solutions to problems and to address needs in society, industry and the environment. The word "technology" can be broadly considered as "know-how" (Phaal et al. 2004)—i.e. the application of scientific and other knowledge in context. Thus, engineering and technological knowledge are closely related, and engineering education includes a combination of "hard" (scientific) and "soft" (craft) knowledge with a practical applied orientation.

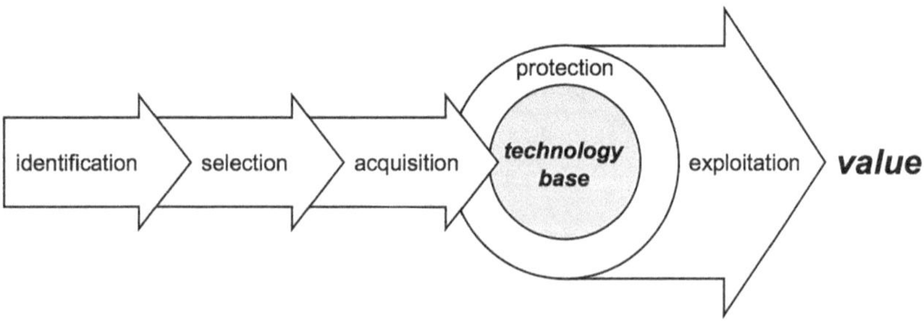

Fig. 1 Technology management process framework. Adapted from Farrukh et al. (2000)

In order to effectively manage engineering and technology, processes are needed to align inputs and activities with desired outputs. The technology management process framework developed by Gregory (1995) is used here, comprising five broad areas of activity:

- *Identification* of technologies that are (or may become) of importance to the business.
- *Selection* of technologies that should be supported by the organisation.
- *Acquisition* of and assimilation of selected technologies.
- *Exploitation* of technologies to generate profit, or other benefits.
- *Protection* of technological knowledge and expertise embedded in products, services and systems.

These process areas are elaborated below (Farrukh et al. 2000), posed as questions that managers, engineers and technologists must address for the effective management of engineering and technological knowledge and resources, from a business perspective.

1.1 How Do We Exploit Our Technology Assets?

In the competitive marketplace, firms that utilise their technological assets most effectively have a significant advantage. Continued exploitation and renewal of the technology base are essential for long-term survival. The systems which support the delivery of products and services to the market need to be clearly understood, in terms of how technology provides value to the company and its customers.

Key issues to consider include: management of the technology base; technology planning, including short- to medium-term forecasts of market requirements and technological capabilities and trends; relationships with the customer–supplier network and with other external sources (for example, standards-making bodies); communication channels and information flows; and operations and resource management.

A clear understanding of the nature of the core technologies in a company is required. How do these relate to the key skills and capabilities of staff, products and services, markets and competitor activity?

The company should be aware of the many options available to exploit its technology, including selling or licensing its technology; joint ventures or collaboration; technology fusion, whereby existing technologies are combined in innovative ways to provide new products or services; technology transfer processes (internal and external); and improved business processes and organisational structures to support the generation and exploitation of technological capability.

1.2 How Do We Identify Technology Which Will Have a Future Impact on Our Business?

Maintenance and renewal of the technology base require that processes are in place for the identification of new technologies which are, or may in the future be, important to the business. This is becoming an increasingly challenging task as the complexity, cost and pace of technological change increase and the sources of technology become more international.

Key issues to consider include: a thorough understanding of the nature of the firm's technology base, in relation to how these add value to its products and services; access to appropriate external and internal networks and sources of information; and knowledge management systems and communication channels, to ensure that the information is appropriately processed and disseminated.

Technology identification processes include a range of activities: systematic scanning of information sources, to develop an awareness of existing and emerging technologies; technology and market foresight and forecasting processes, to support the identification and appraisal of emerging technologies; monitoring of specific technical threats; ways of generating new ideas, to identify new product and process opportunities; technical benchmarking, to develop an awareness of competitors' capabilities; and specific data collection in response to new requirements.

In many organisations, technology identification is undertaken on an ad hoc basis, using informal networks, attendance of trade fairs and conferences, subscription to journals, contacts with suppliers and other means. In addition to making this more systematic, the challenge is to develop appropriate systems to collate and analyse the data collected, and to disseminate it effectively throughout the organisation.

1.3 How Do We Select Technology for Business Benefit?

Managers are commonly faced with difficult decisions about where to invest scarce resources. In the long term, it is critical to select the best technological option, as mistakes can be costly by the time products and services reach the marketplace. Technology investments must lead to increased future revenues and profits which can be re-invested in the technology base for long-term success.

Key issues to consider include: agreement of appropriate decision criteria; establishment of a visible and repeatable decision-making process; understanding the strategic implications of technological choices; and benchmarking with competitors.

Selection of technology is a decision-making process. It requires an understanding of the technology requirements of the organisation, product, service or project, together with the characteristics of identified candidate technologies and any constraints that may affect the selection process. Technology selection involves developing and evaluating alternative solutions, choosing the best option and considering significant implementation factors.

Technology selection decisions can be categorised as being proactive or reactive. Proactive selection decisions are taken in response to future needs, through technology forecasting (future investment in technology for the next generation of products) and technology portfolio analysis (current and future balance of technology). Reactive decisions are taken in response to specific current needs, in relation to current investments (project selection) and urgent problem solving (troubleshooting).

1.4 How Should We Acquire New Technologies?

Organisations need to update and restock their technology base, which can be depleted by obsolescence and diffusion of technology. Specific business reasons for acquiring new technology include: customers in a changing market demand new features in products and services; external constraints (for example, health, safety and environmental legislation) may require the introduction of new products or services; increased competition may demand improvements in technological performance; improved quality requirements may lead to the necessity for upgrading manufacturing and testing equipment; and pressures to reduce costs may require more efficient production processes.

Various routes are available for acquiring new technology, including external purchase or transfer of technology (such as company acquisition, machine purchase or licensing in technology), collaborative development (for example, joint ventures, subcontracting development projects, or supporting supplier technology development) and internal technology transfer or R&D.

Technology acquisition can be seen in terms of a general process: the choice of route for acquiring the technology should be reviewed and assessed; implementation of the chosen route should be managed so that the technology is brought into the organisation to meet required time, budget and performance level requirements; and assimilation of the technology should be achieved to ensure that it becomes a fully accepted and functioning part of the technology base of the company.

1.5 How Can We Protect Our Technology Assets?

A key part of technology management is the maintenance of the technology base. In addition to ensuring that technological resources are renewed, it is important to minimise unplanned transfer of technological assets out of the organisation. Protection of technology involves more than patents and intellectual property rights—it involves people and the knowledge and skills they control, together with other issues such as site and security of information and communications systems.

Key issues to consider include: consideration of the "protectability" of new technologies as part of the selection process; active management of the technology base to ensure awareness of obsolescence and renewal needs; and management of the technological expertise of staff, supported by appropriate reward systems to minimise the risks associated with staff turnover.

Technology protection can involve keeping ahead of competitors by identifying and appropriately securing technology assets (defensive strategy), or by keeping competitors behind by neutralising the effects of their defences (proactive strategy).

Protection of technology should be considered systematically, in terms of a repeating process, with three main stages: assessment of protection need, including a review of existing and new technology assets, in terms of their value to the company (now and in the future); choice of protection routes, based on their suitability for the technology and company; and implementation and enforcement of the protection method.

1.6 Engineering and Technology Management—An Integrated Process View

Effective engineering and technology management requires an integrated approach, so that activities in the five process areas of identification, selection, acquisition, exploitation and protection are aligned, as illustrated in Fig. 1. For reasons of simplicity, the process steps are shown as a rather linear model; in reality, the activities associated with managing engineering and technology are much more diffuse and iterative.

Technology should be considered in the early stages of strategy formulation, and the links with other activities should be clearly understood (such as marketing and other commercial functions, operations, human resources and finance). Mechanisms should be in place to ensure that technology strategies are effectively implemented at the operational level. Technology management processes are often embedded in other business processes. For instance, new technology is often acquired during development projects. It is important to be aware of technology management considerations that continue beyond the completion of the project. The interdependence of technology management processes should be understood, for instance issues of technology protection are an important consideration during technology identification and selection processes.

Engineering and technology management is a very broad subject, and there are many particular areas of importance that require specific attention. For example, innovation, knowledge management, competence and performance measurement are large subjects in themselves. The five-process technology management model presented above provides a framework for understanding how technology can be managed in relation to other relevant management concepts, methods and processes.

2 Case Study

2.1 Identification

This section will use a case study to explore the technology identification tool in terms of technology intelligence. Existing literature argues that technology intelligence has three major questions to address: (i) What do we need to know? (ii) Why do we need to know it? (iii) What decision is to be made, or action taken, once we know it?

In order to address these questions, Kerr et al. (2006) develop a conceptual model for technology intelligence that is concerned with the operating cycle for running a technology intelligence system. The cycle is composed of six phases, namely coordinate, search, filter, analyse, document and disseminate.

Researchers from the Centre of Technology Management (CTM) of the Institute for Manufacturing at the University of Cambridge have visited Xaar plc as part of a project on "Technology Intelligence", and have developed "Xaar Case Study" to provide an insight into the technology intelligence activity within Xaar, and to gain an industrial input for testing CTM's technology intelligence framework that has been mentioned above. In this case study, we limit our discussion on the process cycle of six phases (concerned with the technology intelligence system) that can be demonstrated in the Xaar case.

Xaar plc manufactures and sells high-performance, specialised print-heads and inks to original equipment manufacturers (OEMs) in the graphic arts, packaging printing and industrial printing markets. The company was founded in 1990 to commercially exploit a new digital inkjet printing technology arising out of work done by Cambridge Consultants Ltd. In October 1997, Xaar was listed on the London Stock Exchange. The company's turnover in 2015 was £93.5 Million; the gross investment in R&D was £19.9 Million; and the percentage of gross margin was 47.8%.

From Fig. 2, we can tell that the primary input into the technology intelligence process is the data on technologies from external and internal sources. Adding to this, however, another critical input is also necessary: what information needs to be collected, analysed and disseminated? This means that the second input is the decision-makers' intelligence needs. The output of the technology intelligence activity is the intelligence information for the decision-makers. There are four forms of output: identification of opportunities, awareness of threats, assessment of art and profile of trends.

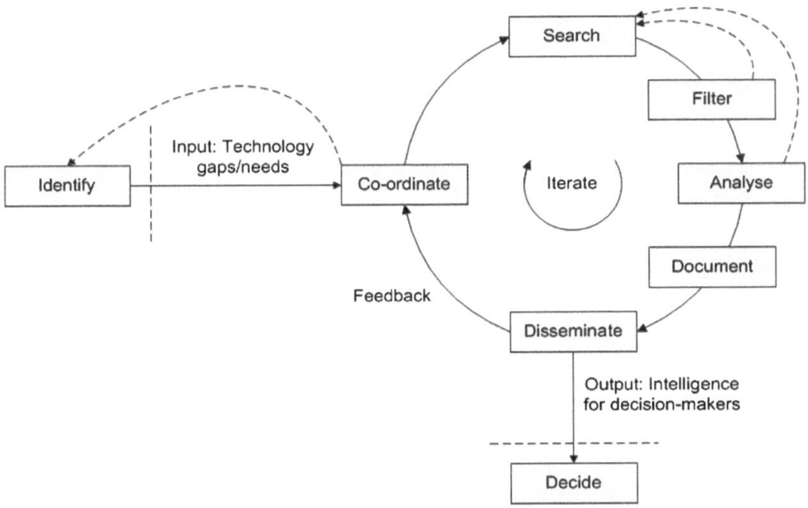

Fig. 2 Technology intelligence's system operating cycle. (Adapted from Kerr et al. 2006)

2.1.1 Coordinate

Consider the operating cycle in Fig. 2; once the input, of needs or requirements, comes into the technology intelligence activity from the intelligence consumers, the first phase is coordinating the technology intelligence efforts needed to fill the gaps in the specific technology know-how. "Coordinate" encompasses the planning of the intelligence activities, allocating resources, briefing agents and gatekeepers, alerting the technology intelligence system to the new intelligence requirements (this involves getting the system sensitive or switching-on the radar to new signals).

In the Xaar case, there exists no formal/structured coordination mechanism for technology intelligence in the sense that the activity and processes must be planned. However, Xaar's Technology Group meets weekly to discuss the company's "Executive Meeting" and this provides the opportunity to delegate specific intelligence projects to individual team members. The group is made up of five–six members who all have a broad range of knowledge, and each individual has two–three technology expertise areas. The lack of a formal coordination mechanism does not appear to cause any "real" problems as the group is small enough for day-to-day interaction. It must be pointed out that there is no form of procedure or checklist of practices. This could potentially mean that an individual on a given occasion could miss or overlook certain aspects of the intelligence activity.

2.1.2 Search

When the activity has been coordinated, the next phase is to "search". This corresponds to the four system modes of mine, trawl, target and scan. Considering searching sources of information, typical sources include: trade shows (direct/personal), patents (direct/impersonal), gatekeepers (indirect/personal), trade journals (indirect/impersonal). In Xaar's case, at the centre of their searching process are three principal sources, namely patents, the Internet and university research centres. There are also many peripheral intelligence sources such as conferences, trade shows, trade magazines and industry bodies, which will not be discussed in detail here.

Xaar has a very strong reliance on patent searching and uses this source to watch targeted companies. They also pay an external provider a subscription fee for access to a commercial inkjet patent analysis report. The company are currently investigating whether to invest in an internal patent database with pre-sorted/pre-filtered material or to commission specific reports from an appropriate external provider. The Internet is used extensively for searching, and Google appears to be the preferred search engine. One of the concerns expressed with this source is the need for validation of the "found" information. Some individuals have addressed this issue by generating their own personal list of reliable bookmarks, for example websites that are tailored to provide quality information in specific technology areas. When using Google, company names are often used as keywords and internally produced company reports published on corporate websites have been found to be good sources of pre-digested material. For certain technology areas, Xaar has links to a number of university research centres and this provides a useful source of intelligence on emerging technologies. However, Xaar does lack a structured approach to search for academic papers published in journals. They are not aware of the academic electronic libraries or bibliographic databases such as ScienceDirect, BIDS or Emerald.

2.1.3 Filter

The search phase is followed by "filter" which determines if the information gathered thus far is pertinent. If it is not, there is a loop back to the search phase for further gathering. As an example, a simple filter could take the form of three stage-gates: (i) Is the information new to me? (ii) Is the information at the correct level and coverage? (iii) Does the information fit to our context/issue?

In Xaar case, the filtering mechanism is effectively left to the individual to use their judgement. There exists no guideline. However, given that the members of the Technology Group are all experienced there is no need to formalise the process. The first filtering decision gate is whether the collected information is new (i.e. not already known); the second gate is whether the information is at the correct level of granularity and with the appropriate coverage needed; and the final decision gate in the internal filtering process is whether the information is fit for Xaar's context. This gate also provides the opportunity to share information, at an early stage of the intelligence activity, by effectively asking another colleague for a second opinion on the relevance/usefulness of the collected information.

2.1.4 Analysis

Filtering is followed by the "Analyse" phase. This is a difficult task involving interpreting the information and relating its relevance to the organisation's particular context and intelligence provision requirements. It reflects the extracting of "value" from the "volume".

Xaar's analysis of the information collected focuses on whether a targeted technology could meet Xaar's intended purpose. This is effectively an analysis for proof of concept. There are two aspects of this analysis—the engineering and the commercial judgements: (i) Is the fundamental technology appropriate? (ii) What would the payback be? Xaar feel that they are very good on making decisions about whether the technology is fit-for-purpose. Initially, the pros and cons for each concept are elicited; the poor technologies are eliminated; the good concepts are then tested in a simulation environment. The commercial perspective is the weakness in the analysis phase. Xaar felt that they were poor at judging commercial readiness of a technology and how much would be generated by incorporating a technology into their portfolio.

2.1.5 Document

"Document" is creating the necessary reporting documentation, structuring the information content of the intelligence and embedding the new knowledge into the organisational memory. This includes information warehousing and knowledge management for accessing and retrieving.

In the Xaar case, the documentation of intelligence findings is ad hoc with individuals left to their own devices to both structure and store the knowledge. There is a shared database within the Technology Group, yet it appears to not be used effectively. A lot of information is stored on team members' individual computers. There are effectively two problems with the documentation phase: (i) a lack of visibility for the intelligence reports and associated ease of locating them; (ii) the form of the information is not readily digestible by other groups. The second problem does form a barrier between the Technology Group and the other departments.

2.1.6 Disseminate

The final phase in the cycle is "Disseminate". This is the trigger mechanism for the intelligence brokers to inform the intelligence consumers to the existence of new/updated intelligence and alerts.

Xaar has a very strong culture of internal networking and "knocking on doors"—"if you want to know something, ask". This is manifested by people coming to desks and having discussions. Internal communication within the Technology Group is very good, and there is the expectation that individuals must proactively share their intelligence findings. This allows team members to gain visibility of their work and ideas. However, cross-group communication is weaker. Some of the interviewees said that information sharing at the company level was poor and as the company grows, it could weaken further. Cross-group exchange is based around "asking". There is no formal forum of "telling" or pushing out technology intelligence findings to other interested parties.

Since the process model is a cycle, there is the option to go around the loop a number of iterations and further refine or tailor the provision of intelligence. Therefore, it can be viewed as a helix process that is continuously refining the interpretation. In a small organisation, a single individual may be responsible for technology intelligence and thus conduct all of the phases (coordinate, search, filter, analyse, document and disseminate), whereas in large organisations, there may be a whole department dedicated to the technology intelligence activity. Thus in various organisational structures, four major intelligence "roles" can be identified: gatekeeper, searcher, technologist and knowledge engineer. The gatekeeper is needed for the coordinate and disseminate phases, whereas the filter and analyse phases require the technologists. The searcher is demanded for the search phase, and knowledge engineer is dedicated for the document phase.

2.2 Selection

Selection of technology is a decision-making process. It involves developing and evaluating alternative solutions, choosing the best option and considering significant implementation factors. We will use the two cases below to explain the selection function in the engineering technology management. Case 1 shows the process of selection of relative prioritisation of R&D projects at BAe, while for the case 2, we develop a technology roadmapping (TRM) framework and use life cycle analysis to select an emerging alternative energy when comparing technology readiness and low-carbon attributes.

2.2.1 Case Study: Relative Prioritisation of R&D Projects at BAe

Research and development within the military aircraft sector of industry is extremely diverse, ranging from short-term demands to satisfy new operational requirements to very long-term ones to meet future defence needs. In the current environment, BAe (like many other companies) cannot satisfy all the R&D that the business demands. It therefore looks towards innovative ways of acquiring the technology that it needs through a mixture of contracts, collaborations and partnerships with both industry and academia, in addition to its own internal R&D programmes.

BAe identified a need to develop an optimum process for the relative valuation of R&D, to enable selection and prioritisation of programmes and give maximum benefit to the military aircraft business (Fig. 3). This would allow the company to make robust decisions on where it should focus its own funding for R&D, both long term and short term, for the benefit of the business. A portfolio approach was developed to represent the cost-to-benefit ratio of each project, together with a measure of customer

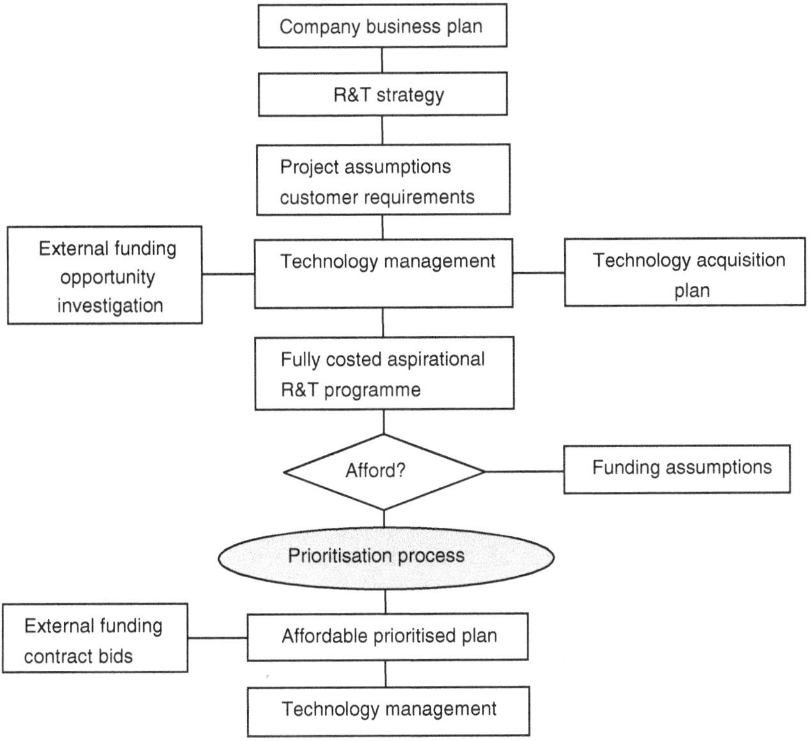

Fig. 3 The process of selection of relative prioritisation of R&D projects at BAe. (Adapted from Venus 1999)

focus. This enables resource allocation decisions to be clearly communicated; the approach has been successfully applied for several years.

2.2.2 Case Study: Roadmapping an Emerging Energy Technology—Dimethyl Ether in China

New energy technologies are becoming increasingly complex. The selection of alterative energies that may replace the existing solutions is strategically important and needs more attention. This case compares two scenarios of dimethyl ether vs diesel and finds that the superiority of dimethyl ether will not arise until 2030, when the complementary engineering technologies become available. We developed a technology roadmapping (TRM) framework to plan and strategise the emergence of a new energy industry that is based on engineering technologies. In order to explore the specific characteristics of an energy sector, we use life cycle analysis to compare both technology readiness and low-carbon attributes of a new and existing energy alternative to have a good selection for the future development.

Dimethyl ether (DME), as an alternative fuel for transportation, has been selected to become the critical case for the examination of the current potential and the future development of strategies/stages of an emerging energy. The effects of two major supporting engineering technologies, such as carbon capture and storage (CCS) and catalytic distillation technology (CDDME), are also carefully examined in this case. DME, given the combustion and auto-ignition characteristics, is an ideal clean-burning substitute for conventional or petrol. Its application helps to reduce harmful emission and alleviate the relieve energy resource shortage (Ou et al. 2010).

In China, domestic research institutions have shown strong interests and carried out many research efforts to understand DME application technology. Several universities, such as Shanghai Jiao Tong University and Tianjin University, have conducted a series of experiments and other research on the DME production process and DME engines, supported by the National Natural Science Foundation or companies like Ford Motors. Since 2007, China has launched a number of DME projects in total exceeding one million tons for operational production. For example, in August 2007, Jiu Tai Energy (Inner Mongolia) Co. Ltd celebrated the ground-breaking ceremony of a 1 million ton/year DME project. A few months later, Shenhua Ningxia Coal Industry Group announced their commissioning in 0.21 million tons DME production.

DME production capacity and output have been growing rapidly recently. In 2001, China's DME production capacity was only 31.8 thousand tons with an output of 20 thousand tons. And by 2006, the two numbers were increased to 480 thousand tons and 320 thousand tons, respectively, with annual growth rates of 97% and 96%. By 2008, there were 52 DME producers in China with existing capacity of 4.18 million tons and the capacity would reach 15.8 million tons in 2012. However, there are several main growth barriers to coal-based DME, such as high carbon emissions in the fuel stage, high production cost and low energy efficiency.

2.2.3 DME versus Diesel in Two Scenarios

Life cycle analysis will help us to plan the growth of DME when assessing the key obstacles to DME through the entire life cycle: environment impacts, economics value and energy consumption. The TRM tool, in turn, will help us to plan when and why we should select DME over diesel as a transportation fuel. As per the recommendation of the experts, two scenarios have been planned for further analysis: (i) DME versus diesel in 2020; (ii) DME versus diesel in 2030.

The use of the framework for roadmapping the emergence of the new technology is illustrated below by means of a case study, focusing on the comparison between DME and diesel. There are two scenarios analysed through LCA method, in order to explore and clarify the unknowns in an a priori framework. The development strategies and key stages have been clarified in the refined framework, and the supporting policies are subsequently suggested.

The case study aims to plan the development strategies and key stages of DME in China. Some basic assumptions have been collated as follows (Zhou et al. 2012):

- Huge market demand: By 2030, China may need to import oil for 800 million tons per year. It is urgent to find an alternative fuel (i.e. DME) for transportation to replace diesel or petrol in China.
- Complementary technologies to DME: CCS technology will develop at a fast pace. It will have market demonstration in 2020 and will be implemented in 2030. CDDME technology is in the tech-demonstration stage now and will be implemented in 2020.
- Coal production in China: In 2030–2050, the production will be 3.8 billion tons per year.

2.2.4 Scenario I: DME vs. diesel in 2020

The required experts' estimations and assumptions have been collated as follows, supported by documentary data:

- Energy consumption on production process: In 2020, CDDME technology is still in its embryonic stage, and its efficiency still remained low. In the "fuel stage" of the life cycle analysis, coal-based DME needs five times the consumption during the production stage than ordinary diesel (from crude oil).
- Carbon emission: In 2020, CCS technology is still in its infancy for technological demonstrations. Therefore, the carbon emission of DME production will be two times of diesel.
- DME production cost: DME is still in its embryonic stage, so the price in 2020 might be USD 220 per barrel.

From Table 1, we argue that DME will not be able to challenge diesel in 2020, as its key performance indicators are significantly inferior to those of diesel.

2.2.5 Scenario II: DME vs. diesel in 2030

- Energy consumption in the production process: In 2030, CDDME technology may have been significantly refined in terms of efficiency. In the "fuel stage" of the life cycle analysis, coal-based DME requires three times the energy consumed during the production stage than that of ordinary diesel (from oil).
- Carbon emission: In 2030, CCS technology may have been its early stage of implementation. Assuming 45% efficiency, the carbon emission of DME production will be 1.08 times of diesel.
- DME production cost: Including the carbon trading gain (benefit from CCS), DME might have the price of USD 135 per barrel.

Table 1 LCA analysis: DME versus diesel in 2020. Adapted from Zhou et al. (2012)

DME versus diesel	Fuel stock	Fuel	Vehicle	Total
Energy consumption	0.45:1	5:1	0.68:1	1.47:1
Price	N.A	N.A	2.2:1	2.2:1
Carbon emission	1.76:1	1.53:1	0.8:1	2:1

Table 2 LCA analysis: DME versus diesel in 2030. Adapted from Zhou et al. (2012)

DME versus diesel	Fuel stock	Fuel	Vehicle	Total
Energy consumption	0.45:1	3:1	0.68:1	0.981:1
Price	N.A	N.A	1.35:1	1.35:1
Carbon emission	1.13:1	1.2:1	0.8:1	1.08:1

From Table 2, we argue that DME will start to be able to challenge diesel in 2030, as its key performance indicators are almost on par with those of diesel. The industrial strategy would need to expand the supply and penetrate the mass market.

Through the selection and comparing of DME versus diesel, we can find that, in 2020, policy should consist predominantly of supply policies, such as giving R&D grants, encouraging its application and demonstration, etc. In 2030, policy should be more market and environment oriented, such as industrial standards, regulations and stipulations, and application networks should be supported.

2.3 Acquisition

Organisations need to update and restock their technology base, which can be obsolescence and diffusion of technology. Various methods are available for acquiring new technology, including technology transfer, such as company acquisition, machine purchase or licensing in technology; collaborative development, for example joint ventures, subcontracting development and so on. We use two cases to illustrate technology acquisition. The first one is a successful purchase of the company assets of Domino. The second is about firm-level technology transfer and technology cooperation for wind energy between Europe and China.

2.3.1 Case Study: Acquiring a Differentiating Technology for a New Range of Products at Domino

Satisfying customer requirements for improved products is a key driver for the acquisition of new technologies in manufacturing companies.

Domino Printing Sciences has been at the forefront of inkjet printing technology for marking and coding systems for many years. The requirement from customers for cleaner, more reliable coding technologies encouraged Domino to investigate potential alternatives to serve existing markets and to open up new market possibilities.

A systematic review of coding technologies was undertaken, leading to the identification of lasers as a cleaner alternative. Lasers can mark many materials directly, such as plastics or glass, where surface discoloration acts as a mark. To survey the laser marketplace, guidance was sought from a technical consultancy.

Domino had a clear strategy for acquiring laser technology. As the laser would be the main differentiating element in a product coding system, Domino required complete control over the design and manufacture of the laser. Lasers were too expensive and not sufficiently developed for this application to buy in ready-made. A programme of R&D to produce a low-cost, reliable laser development partner could be found.

Domino identified a company in the USA which had laser design and manufacturing skills and which had developed a unique, fast and robust marking product. This company had good laser technology, technologists and facilities, but had suffered from poor marketing and was not profitable. Such a company would provide Domino with the technical laser capability it needed to integrate into its new generation of marking system. A successful purchase of the company assets was made. A critical condition was the retaining of a few key specialists.

In addition to this, a key customer had identified a small laser-making company in the UK that had developed a high-resolution laser market. This company was looking for a bigger company to work with in developing its product. Domino acquired the US laser company at the same time developing an exclusive licensing partnership with this UK company. The high-resolution marker was complementary to the US product but could use the same laser technology.

There was initial concern that laser products would compete with existing inkjet technology and replace this element of the business. However, it has been realised that there was a new market waiting in anticipation for the new product and that, far from competing, both technologies have complemented each other. Domino expects to be world market leader in laser marking within 2 years.

2.3.2 Case Study: Firm-Level Technology Transfer and Technology Cooperation for Wind Energy Between Europe and China

Technology transfer and cooperation as the tools of technology acquisition are key mechanisms for transferring low-carbon innovation from high-income countries to low- and middle-income countries. In this case, we try to explore how and to what extent technology transfer and cooperation from the EU have shaped the leading firm-level wind energy technologies of China today. China is the world's largest wind energy market, and four of its biggest wind energy firms: Goldwind, Sinovel, Guodian United Power, and Mingyang are part of the global top 10. Technology acquisition plays an important role in promoting the development of wind energy in China. For example, Lema et al. (2015) argue that there is a relationship between 26 Chinese turbine manufacturers and 18 (mainly) European knowledge-intensive businesses, most of which are German.

Table 3 indicates the relationship between Chinese wind energy firms and European wind energy firms with regard to different models of technology transfer and technology cooperation. Interesting to note is that the four top Chinese wind firms (indicated in cursive in the table) have all built their wind energy expertise on technology transfer from European, mostly German, wind energy firms: Goldwind has conducted joint development with Vensys as well as licensing from Jacobs/REpower, Guodian United has licensed technology from Aerodyn, Mingyang had joint development with Aerodyn, and Sinovel had licensed technology from Fuhrländer.

Table 3 The relationship between Chinese and European wind energy firms with regard to different models of technology transfer and technology cooperation. Adapted from Lewis (2013)

Chinese company	Model of technology transfer/cooperation	European source firm
A-Power (GaoKe)	Licence	Fuhrländer
	Licence/joint development	Norwin
Beijing Beizhong	Licence	DeWind
CSIC Haizhuang	Licence	Frisia
	Joint development	Aerodyn
DEC	Licence	REpower
	Joint development	Aerodyn
Goldwind	Licence	Jacobs/REpower
	Joint venture/acquisition	Vensys
Guodian United Power	Licence	Aerodyn
Harbin Stream Turbine Co.	Licence	Aerodyn
Hewind	Joint development	Aerodyn
Huidde	Licence	Fuhrländer
Jiuhe	Licence	Windrad Engineering
Mingyang	Joint development	Aerodyn
REpower North	Joint venture	REpower
Sewind		DeWind
	Joint development	Aerodyn
Sinovel	Licence	Fuhrländer
Windey	Licence	REpower
Xi'an Aero Engine Corp.	Joint venture	Nordex
Xi'an-Nordex	Joint venture	Nordex
Yinhe Avantis	Joint development	Avantis energy

2.3.3 Technology Cooperation in Europe and China: Vensys–Goldwind Vensys

Vensys is a German wind turbine manufacturer that was acquired by the Chinese firm Goldwind in 2008. Vensys started as a small engineering bureau that emerged from an R&D centre at the University of Saarbrücken. Vensys has been commercially operating in Germany since 2000, whereas the R&D activities at the university started about 10 years earlier. Vensys was acquired by the Chinese wind firm Goldwind with a share of 70% in 2008. Vensys operates via licensing in China (Goldwind), India (ReGen Powertech), Brazil (Enerwind/IMPSA wind) and Spain (EOZEN). It has strict rules for licensing to ensure that its intellectual property rights (IPRs) are protected. It sells its turbines in Brazil, Bulgaria, Canada, China, Germany, India, Pakistan, Poland, Portugal, Russia, Romania and the USA. Goldwind provided Vensys with access to the Chinese market and contacts; it enabled small firms to upscale rapidly and to supply a huge market. Goldwind has access to Vensys' profits, technology, IPRs, its components and markets. Vensys is famous for developing the permanent magnet direct-drive (PMDD) which is a technology based on a permanent magnet that powers the drive, hence different from the electromagnetic direct-drive Enercon uses. The acquisition of Vensys by Goldwind has contributed to the internationalisation of EU wind markets and technology. German and Chinese technology cooperation has led to joint R&D and joint technology. Vensys' PMDD technology requires the use of rare earths. The PMDD fits very well for production in China since China is one of the few countries that have access to rare earths resources, whereas other countries—such as Germany—struggle to access rare earths (Nordensvard and Urban 2015; Lema et al. 2015). It sells medium to large turbines and is currently conducting R&D for a 10 MW turbine.

Goldwind was founded in 1998 and is headquartered in Xinjiang. As one of the earliest wind energy firms in China, it evolved in many parts of the wind energy business, including wind turbine design and manufacturing, wind resource assessment and wind farm operation. In recent years, Goldwind has become the largest manufacturer of wind turbines in China and the second largest globally (CIEDS 2013). It has a market share of about 20% and is said to have installed a generating capacity of about 3600 MW (Li et al. 2013). With strong, internationalised R&D capabilities, Goldwind has become the world's largest manufacturer of PMDD wind turbines. For now, Goldwind has its branches and factories located in six continents.

Goldwind experienced several key innovation paths along its development. Goldwind started the development and marketing of 600 kW and 750 kW in the 1980s, leading the Chinese wind market. The early turbines installed in China relied on imported components from

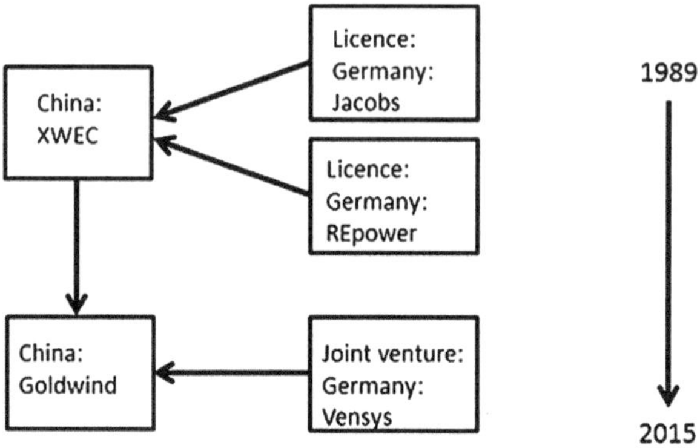

Fig. 4 Goldwind's technology cooperation with Germany from Jacobs, REpower and Vensys. Adapted from Lewis (2013)

technology transfer. Already in 1989, the predecessor of Goldwind, China XWEC, licensed wind energy technology from German wind firm Jacobs Energie (see Fig. 4). China's first five turbines installed in 1998 had only about 33% local content. In 2001, Jacobs Energie merged with another company to form the REpower Systems Group. That same year, Goldwind obtained a licence from REpower for a 750 kW turbine. In both cases, Goldwind insisted to add technician and researcher training in the contract. While Chinese engineers were sent to Germany for operational training, experts from Jacobs and REpower also went to China to work and provide on-site training. Through the immersion of design teams and experimental learning processes, Goldwind improved its innovation capacity and successfully produced turbines of 600 kW and 750 kW in 1999 and 2001, respectively. This forms the bases for later joint research of the 1.2 MW turbine with Vensys.

In 2003, Goldwind embarked on the collaborative design of a 1.2 MW PMDD wind turbines with Vensys. Unlike REpower, Vensys was a design firm who therefore was complementary to a manufacturer like Goldwind. However, Vensys only designed gearless turbine technology (direct-drive gearless wind turbine), which was uncommon back then and is different from Goldwind's previous innovation paths—Goldwind produced turbines with gears, namely doubly fed induction generators of 600 and 750 kW before. Advantages in the new innovation path meant that gearless turbines had less weight, less cost, less parts for maintenance and replacement. When considering the strategic potential, Goldwind determined to take the risk and commit to this new technology. In 2005, Goldwind had the prototype of the 1.2 MW turbine and installed it in Da Ban City wind farm for pilot operation. That became the first wind turbine produced in China over 1 MW.

Furthermore, Goldwind also acquired the licence for that 1.5 MW turbine with a larger 64-metre-diameter rotor, when integrating knowledge from Vensys to R&D teams in China (Lewis 2013). Based on this, Goldwind improved its magnetic electric direct-drive technology to produce 1.5 MW turbines in 2007. After its acquisition of Vensys in 2008, Goldwind had already commercialised the products of the 2.5 MW (2009) and 3.0 MW (2009) turbines, through internalising Vensys' R&D competences. In summarising key factors for the success of Goldwind, former CEO Wu Gang emphasised that "insisting on collaborative research, rather than licensing technology or purchasing turbine design solutions made Goldwind strong at independent technology development" (interview 2010). After acquiring Vensys by 70% in 2008, the registered patents for Goldwind increased from 3 in 2007 to over 170 in 2012 (Zhou et al. 2015).

Goldwind then established a joint venture with Vensys for developing 1.5 MW and 2.5 MW direct-drive wind turbines, which made up around 20% of the total production capacity in 2012 (Urban and Zhou 2015). These wind turbine models are estimated to dominate the majority of the wind market in China for the next 3–5 years, according to expert views. After executing its internationalisation strategy, Goldwind is developing key products for the future, including wind turbines in the size of 6.0 MW to 10 MW for offshore use.

In addition, Goldwind and Vensys are conducting joint R&D on amending turbines for the local conditions in China. This requires turbines that are suitable for low wind speed areas and extreme conditions such as desert conditions involving high heat, extreme dry weather and extreme sand exposure (e.g. in Gobi desert) and high altitude (e.g. for the Tibetan plateau).

Technology acquisitions in this case have two main implications. First, technology acquisition has a great effect on "catch-up" countries' innovation trajectories. Asian innovation paths in wind energy, particularly in China, have to some extent evolved based on the technology they acquired from their European technology cooperation partners. Wind energy technology from Europe has therefore helped shape Chinese wind energy technology. Second, technology acquisition provides an opportunity for the small firms in developed countries. The entry of Asian wind energy firms into European markets as well as the entry of European wind technology in Asian markets has led to an internationalisation of global wind energy markets and technologies. Large Asian wind energy firms such as Goldwind offer opportunities for profits, employment and economic growth for smaller wind design firms such as Vensys.

2.4 Exploitation

Exploiting technology assets involves a clear understanding of the nature of the core technologies and opportunities in a company, management of the technology base, technology planning and relationships with the customer–supplier network and other external resources. Hence, we will adopt the two cases GEC-Marconi and digital camera to explore this. For the GEC-Marconi case, which will help us to have a better understanding on how to exploit synergies between the various operating units and sharing of resources in key areas and improving technology planning in the context of the business/marketing objectives of the firm, while the case of digital camera develops a roadmapping method to explore the nature of a potential future value opportunity and articulate the route towards successful exploitations.

2.4.1 Case Study: Exploitation of Cross-Business Technology Synergies at GEC-Marconi

GEC-Marconi is a large international multi-business corporation with a turnover of over £3 billion. The company produces high-technology, electronics-based products for a large number of applications in a wide variety of military and commercial markets. A range of concurrent technology planning initiatives is being undertaken within the organisation, with the following aims: improving the exploitation of the technological synergies between the various operating units and sharing of resources in key areas; improving technology planning in the context of the business/marketing objectives of the firm and more closely integrating the role of central R&D facilities.

As part of this process, a simple matrix-based method was used to develop a framework to link technological capabilities with business objectives. This involved segmentation of the business in terms of technology and business areas in a series of senior management workshops. By ranking and assessing the impact of each technology area on each business area, it was possible to identify core technology areas which are of high value across several business units and areas of mismatch between value, effort and risk. This has enabled the organisation to focus attention on, and investment in, key areas of common interest and to achieve greater levels of coordination between historically independent business units.

2.4.2 Case Study: Charting Exploitation Strategies for Emerging Technology

Exploitation in emerging technology is a risky business, but it is crucial for a firm to achieve future economic prosperity. Continued exploitation and renewal of the technology base are essential for a long-term survival. Emergence roadmapping (ERM) is a workshop method that supports rapid strategic appraisal of early-stage technologies for the exploitation. The approach, which is based on earlier work demonstrating patterns in the historical emergence of industries (Phaal et al. 2011), has been developed and tested in collaboration with technology ventures, established businesses and academic research groups.

The ERM method follows on from the value roadmapping (VRM) approach (Dissel et al. 2006), which enables value opportunities for emerging technology to be identified and prioritised. The ERM method provides a structured process for these opportunities to be explored further, to clarify the strategic direction and to agree on technical and business development actions necessary to move forward.

2.5 The Case of the Digital Camera

The emergence of consumer digital cameras, from initial developments in the 1960s through the development of a mass consumer market in the 1990s, provides an illustration of the patterns governing the emergence of early-stage technologies (see Fig. 5).

Key milestones in this journey were:

- A 1961 paper from the Jet Propulsion Laboratory described the concept of using mosaic photosensors to produce still digital images, which led to the invention of the charge-coupled device (CCD) at the AT&T Bell Laboratories in 1969 (applied science demonstration).
- The technology was first commercialised by Fairchild Semiconductors and was rapidly incorporated into a prototype camera system by Kodak in 1975 (technology demonstration).

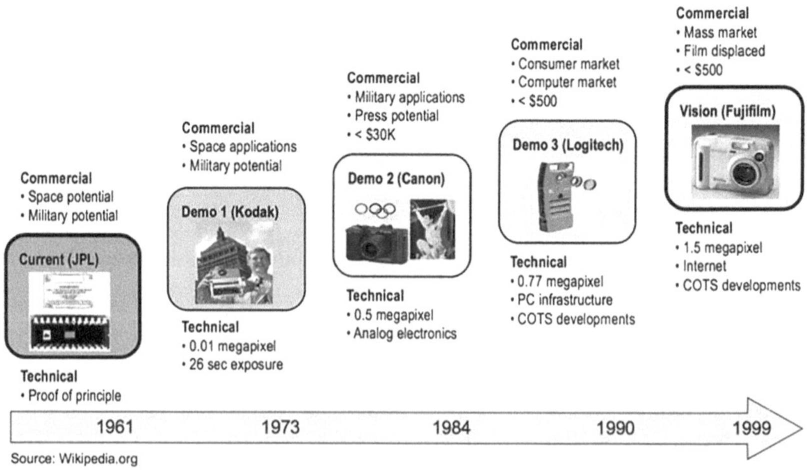

Fig. 5 Key demonstrations milestones in the emergence of consumer digital cameras. Adapted from Phaal et al. (2012a, b)

- Space and military applications enabled the technology to be improved, and the price reduced until eventually the first "mass" market (professional press) was stimulated by a demonstration of the technology by Canon—an image taken at the 1984 Olympic Games in Los Angeles was transmitted and printed in a Tokyo newspaper (application demonstration).
- The price–performance ratio of the core technology continued to improve, with parallel developments in electronics, software and computing supporting the core technological developments, leading to the first consumer digital camera product being released by Logitech in 1990 (commercial application demonstration).
- Sensor technology continued to develop, along with complementary developments in computing, communications, standards, displays, batteries, and printing and scanning systems. This led to cameras that could compete with, and eventually displace, film-based technology, typified by the Fujifilm MX-600, released in 1999, which offered all of the main features that are expected in compact consumer cameras today (price–performance demonstration).

Emergence roadmapping can facilitate the decision-making progress for early-stage technologies by allowing workshop participants to rapidly map the potential commercial exploitation paths for a technology on the industrial emergence framework, tracing its potential trajectories through a series of demonstrator steps.

The aims of the ERM workshop are:

- To clarify the innovation opportunity, in terms of application, market and technology;
- To define steps towards the opportunity, mapping the demonstration chain;
- To explore key enablers and barriers as well as next actions to move towards the first demonstrator.

The journey from science to mass consumer market was long, although there were opportunities to generate revenue earlier in specialised precursor and embryonic markets. In the 1970s and 1980s, products were developed for space, military and professional press markets, each of which would have its own science–technology–application–market life cycle. In the context of the consumer digital camera industry, these achievements can be considered as application demonstrations that enabled continued improvements in the performance of the technology and reductions in its cost.

The historical route from science to mass market application for digital cameras is clear in hindsight. Of course, the future is less predictable, and it would have been unreasonable to expect anyone in 1961 to have foreseen the key developments that would lead to mass commercialisation of this technology. However, investments must still be made, and it is necessary to imagine and explore potential future value opportunities in order to build confidence about the decisions and actions required to move forward.

The workshop allows a detailed exploration of the opportunity, the different stages of its progression towards the ultimate goal, considerations of who and what should be involved along with internal and external factors that may help or hinder progress, and associated actions. The approach requires a relatively clear focus in terms of potential future value opportunity scenarios, including application and market. The pattern of emergence typified by the development of the digital camera industry offers a framework within which to consider these high-risk decisions and make a sustainable development of the company.

2.6 Protection

As aforementioned, protection of technology involves more than the legal rights of intellectual property—it involves people and the knowledge and skills they control, together with other issues such as site and security of information and communications systems. Technology protection can involve keeping ahead of competitors by identifying and appropriately securing technology assets (defensive strategy), or by keeping competitors behind by neutralising the effects of their defences (proactive strategy). Protection should be considered systematically in a circular process. The three main stages involve: the assessment of protection that are needed, the choice of protection routes or mechanisms, and the implementation of the protection methods. Following two cases are discussed: case 1 shows the way of protection method, and case 2 demonstrates the differences of knowledge sharing in open-source innovation as well as the geographical differences between the East and the West.

2.6.1 Case Study: BG plc

In February 1997, British Gas demerged its gas trading and associated activities and renamed itself BG plc. The new company, Centrica, uses the trading name of British Gas in the UK. BG plc uses British Gas outside the UK, which was reorganised in 1999 as BG Group plc. BG plc has operations in 25 countries across Africa, Asia, Australasia, Europe, North America and South America and produces around 680,000 barrels of oil equivalent per day. It has a major liquefied natural gas (LNG) business and is the largest supplier of LNG to the USA. On 31 December 2009, it had total proven commercial reserves of 2.6 billion barrels (410,000,000 m^3) of oil equivalent. BG Group is listed on the London Stock Exchange; as of 6 July 2012 it had a market capitalisation of £44.9 billion.

Organisations such as BG plc are recognising that a significant part of their value lies in the knowledge which the company and its employees possess, rather than just in its physical assets. This is particularly true for knowledge of technology. The effective management of that knowledge can lead to an enhancement in the performance of a company. However, this can only be achieved by a change in culture and working methods, whereby the creation and sharing of knowledge is both encouraged and rewarded.

The mission of the knowledge management technologies team within BG technology is to contribute to and to facilitate the optimisation of the use of BG's world-leading knowledge base in all aspects of gas technology. The approach has been to work with the BG business to define where knowledge sharing, particularly in the field of technology, can enhance their performance, and then to design and deliver information systems which can realise that potential enhancement.

The development of a technology bank is an example of such an activity. The technology bank is maintained by programme managers within BG technology. Current and past technology knowledge is captured in databases within the bank. This then provides BG with the ability to share this knowledge worldwide between virtual teams. Highly advanced tools have also been developed to search for information simultaneously across all of these and many other databases. Information in the databases can be made visible to everybody within the company or only to restricted groups of people, depending on its level of confidentiality. It is planned to extend this capability to the sharing of selected information with partners outside the company.

This project will involve users from around the world and will also involve changing the way of working within BG technology, so that as project information is created it is automatically captured within the databases. The project is just one example of BG's move towards becoming a knowledge-based company, enabling all the knowledge to be accessed quickly and easily from anywhere in the world.

2.6.2 Case Study: Knowledge Sharing in Open-Source Software Projects

Open-source software (OSS) is a software with its source code available with a licence in which the copyright holder provides the rights to study, change and distribute the software to anyone and for any purpose. The Linux operating system, Mozilla and Chrome are all outstanding representatives of OSS, which is widely used all over the world.

OSS projects aim to develop OSS by means of groups of capable people, mainly including developers and users. In terms of developers, core members, active members and peripheral members are three major groups. For users, both developer-users and non-developer-users, their feedback is of great importance for developing work. As distinct from proprietary software with "the cathedral model", OSS works on "the bazaar model" in which everyone can get involved or leave at any time (Table 4) and any one has equal rights to contribute (Raymon 2001).

Table 4 Difference between the cathedral model and bazaar model. Adapted from Panchal and Fathianathan (2008)

Factor	Cathedral model (traditional collaboration)	Bazaar model (distributed innovation)
Structure	Hierarchical	Flat network
Participants	Task oriented	Interest oriented
Production	Management by bureaucracy	Management by objective
Division of work	Distributed by leaders	Self-control
Knowledge flow	Top-down, bottom-up	Distributed
Release	Release after final revision	Continuous revision and release
Decision	Concentrated	De-centralised

Many countries are paying much attention to the development of open source. For example, on 23 May 2012, President Obama issued a directive entitled "Building a 21st Century Digital Government" to promote open government, open data and an open-source plan. OSS is regarded as a way of making it easier for the government to share data, improve tools and services, and return value to taxpayers (Obama 2012). Taking the Columbia government of United States as an example, the government organised an open-source competition named "Apps for Democracy" for the public, and there were 47 pieces of OSS projects created within 30 days to improve government decisions (Booth 2010). The Chinese government continuously take measures to promote development of open source. On 16 March 2011, the Chinese central government issued the 12th Five-Year Plan to support the construction of the OSS ecosystem. Then, on 21 March, the OSS Innovative Lab was established with the sponsorship and collaboration of the Ministry of Industry and Information Technology, the National University of Defence Technology and the International OSS Community—Ubuntu. On 8 April 2014, the Ministry of Industry and Information Technology of the People's Republic of China claimed to support R&D and implementation of the Linux operating system after Microsoft stopped services to Windows XP users. On 16 May 2014, central governments were forbidden to install Windows 8 operating system during the Government Buying owing to the consideration of government information security. To protect government information security, it is considered more and more important to promote the OSS ecosystem.

Open-source ecosystem can be compared in view of OSS project ecological environment, and the differences between China and the West were found to be as follows:

- Different internationalisation levels caused by various languages, political and cultural backgrounds. OSS project is generally communicated and recorded in English, which forms the barriers to most Chinese developers.
- Differences between the spirit of open source and the ideology of sharing owing to the different stages of open source. Developed countries have more experience of the open-source area, while China has much room to improve concerned with the awareness of open-source spirits as well as cooperation and sharing.
- Different occupation systems and habits caused by the influence of both the values of "official standard" and "technology first". For a long time, China has been influenced by the idea of an "official standard" based on official orientation, authority and respect. So few Chinese developers remain to work as technical experts upon 35 years old, while in developed countries there are many savvy technical experts with the age of 40–45 years old.
- Different extent of support for OSS projects. In developed countries, big companies or foundations are supporting many open-source projects overseas, such as Linux, Apache, FreeBSD and Debian; however, in China, the support is limited.
- Differences in intellectual property. OSS licences are very important in OSS, which seems like the Code used in OSS. Undoubtedly, Western countries pay close attention to the licence selection and use of OSS, as deep-rooted concepts exist regarding the protection of intellectual property rights. By contrast, the environment for intellectual property is relatively poor in China

In the development of OSS, the West pays more attention to industrial orientation. That is to say, big companies, foundations and civil senior open-source people contribute a lot, but the government's role remains limited. On the contrary, Chinese government is more proactive to address the need "to support the development and application of OSS, and quicken the formation of an industrial ecological system based on the open source mode" in the 12th Five-Year Plan of Software and Information Technology Service Industry. However, at present, there are no obvious effects; for example, pirated software runs wild, and intellectual property is ineffectively protected. Therefore, in the field of OSS ecological environment construction, the government should pay attention on stressing the protection of intellectual property, formulating fair and reasonable rules, promoting the completeness of the relevant laws and regulations in China, as well as encouraging government procurement to cultivate the domestic market. University–industry collaboration with regard to OSS should also be strengthened.

References

Booth, D. R. (2010). *Peer participation and software: What Mozilla has to teach government*. Boston: The MIT Press.

Bower, J. J., & Christensen, C. M. (1995). Disruptive technologies: Catching the wave. *Harvard Business Review*, 43–53.

CIEDS Chinese Institute of Engineering Development Strategy. (2013). *Report on the development of China's strategic emerging industries*. Beijing: Science Press.

Dissel, M. C., Phaal, R., Farrukh, C. J., & Probert, D. R. (2006). Value roadmapping: A structured approach for early stage technology investment decisions. In *Technology Management for the Global Future, PICMET, 3*, 1488–1495.

Farrukh, C. J. P., Phaal, R., & Probert, D. R. (2000). *Technology management assessment procedure – A guide for supporting technology management in business*. London: Institution of Electrical Engineers.

Frey, C. B., & Osborne, M. A. (2013). The future of employment: How susceptible are jobs to computerisation? *Oxford Martin School (OMS)* (Working Paper). University of Oxford. Retrieved October 9, 2016, from http://www.oxfordmartin.ox.ac.uk/downloads/academic/The_Future_of_Employment.pdf.

Gregory, M. J. (1995). Technology management: A process approach. *Proceedings of the Institution of Mechanical Engineers*, 209, 347–356.
Kerr, C., Mortara, L., Phaal, R., & Probert, D. R. (2006). A conceptual model for technology intelligence. *International Journal of Technology Intelligence and Planning, 2*(1), 73–93.
Lema, R., Nodensvard, J., Urban, F., & Lutkenhorst, W. (2015). *Technological trajectories for climate change mitigation in Europe, China and India, Wind energy in Europe* (*GDI* Working Paper). Bonn: GDI.
Lewis, J. (2013). *Green innovation in China: China's wind power industry and the global transition to a low carbon economy*. New York: Columbia University Press.
Li, J., et al. (2013). *China wind power outlook 2013*. Beijing: China Environmental Science Press.
Mitcham, C. (1978). Types of technology. *Research in Philosophy & Technology, 1*, 229–294.
Obama, B. (2012). *Digital government: Building a 21st century Platform to better serve the American people*. The White House.
Ou, X., Yan, X., & Zhang, X. (2010). Using coal for transportation in China: Life cycle GHG of coal-based fuel and electric vehicle, and policy implications. *International Journal of Greenhouse Gas Control, 4*(5), 878–887.
Phaal, R., Farrukh, C. J. P., & Probert, D. R. (2004). A framework for supporting the management of technological knowledge. *International Journal of Technology Management, 27*(1), 1–15.
Phaal, R., O'Sullivan, E., Routley, M., Ford, S., & Probert, D. (2011). A framework for mapping industrial emergence. *Technology Forecasting and Social Change, 78*(2), 217–230.
Phaal, R., Kerr, C., Oughton, D. & Probert, D. (2012a). Towards a modular toolkit for strategic technology management. *International Journal of Technology Intelligence and Planning, 8*(2), 161–181.
Phaal, R., Routley, M., Athanassopoulou, N. & Probert, D. (2012b). Charting exploitation strategies for emerging technology. *Research-Technology Management, 55*(2), 34–42.
Panchal, J. H. & Fathianathan, M. (2008). Product realization in the age of mass collaboration. In *ASME international design engineering technical conferences and computers and information in engineering conference* change (Vol. 1, No. 34, pp. 219–229).
Raymond, E. (2001). *The cathedral & the bazar-musings on Linux and open source by an accidental revolutionary*. O'Reilly Media.

Urban, F., Nordensvard, N., Nordensvard, J., & Urban, F. (2015). The stuttering energy transition in Germany: Wind energy policy and feed-in-traffic lock-in. *Energy Policy, 82,* 156–165.
Urban F., Zhou, Y. Nordensvard, J, & Narain, A. (2015). Firm-level technology transfer and technology cooperation for wind energy between Europe, China and India: From North-South to South-North cooperations? *Energy for Sustainable Development*, 28, 29–40.
Venus, T. (1999). Getting value from technology: A guide to technology management. Cambridge: University of Cambridge Centre for Technology Management.
Zhou, Y., Li, X., Lema, R., & Urban, F. (2015). Comparing the knowledge bases of wind turbine firms in Asia and Europe: Patent trajectories, networks, and globalization. Science and Public Policy. Retrieved October 10, 2016, from http://spp.oxfordjournals.org/content/early/2015/09/29/scipol.scv055.full.
Zhou, Y., Xu, G., Zhi, Q., Zhang, F., Minshall, T., & Su, J. (2012). Roadmapping an emerging energy technology: Annex-ante examination of dimethyl ether development in China. *International Journal of Product Development, 17*(3–4), 296–310.

Authors' Biography

Yuan Zhou (Joseph) is an Associate Professor at School of Public Policy and Management, Tsinghua University (China). He received his PhD degree (technology management) from the University of Cambridge (UK) in 2010. Zhou Yuan holds B.Eng. and M.Sc. degree from Nanyang Technological University.

Tim Minshall is Reader in Technology and Innovation Management at the University of Cambridge, and Head of the Centre for Technology Management in the Department of Engineering. Academic interests include open innovation, university–industry collaborations, technology business investment and incubation, engineering education and digital fabrication.

David Probert is Reader in Technology Management, founding member and emeritus Head of the Centre for Technology Management within the Institute for Manufacturing, at the University of Cambridge Engineering Department. This is a team of around 40 research and academic staff, with an active consortium of collaborating companies.

3

International Manufacturing and Engineering

Yongjiang Shi and Yufeng Zhang

1 Introduction

The manufacturing system has evolved into various kinds of network-based relationships from the traditional input–output transformation model. During the last 20 years multinational corporations (MNCs) have attempted to globalise their geographically dispersed factories by coordinating them into a synergetic network (Flaherty 1986; Ferdows 1997; Shi and Gregory 1998). This transformation has changed basic manufacturing functions and effectiveness from the orientation of product-based competitive advantages towards the orientation of network strategic capability development, which drives the manufacturing system beyond a factory and the strategy beyond product focus.

Y. Shi (✉)
Institute for Manufacturing, Cambridge University, Cambridge, UK
e-mail: ys@eng.cam.ac.uk

Y. Zhang
Birmingham Business School, University of Birmingham, Birmingham, UK
e-mail: zhangys@bham.ac.uk

Y. Zhang and M. Gregory (eds.), *Value Creation through Engineering Excellence*,
https://doi.org/10.1007/978-3-319-56336-7_3

Besides MNCs' international expansions, it has become more popular for all types of companies to downsize and outsource their non-core business tasks and to set up inter-firm collaborations (Lambert et al. 1998; Lamming et al. 2000; Brewer et al. 2001). This development has pushed the manufacturing system further into a new relationship beyond the traditional concept of a firm that owns and internally operates its factories. Currently, it is no longer a secret that, although a firm may only own a very small portion of a supply chain, they are still strategically able to coordinate or integrate the whole supply chain to deliver a competitive product to its targeted market. It is equally interesting to notice that there are increasing observations about geographic clustering emerged worldwide (Piroe and Sabel 1984; Porter 1998). The clusters actually form different supply networks—some of them are internally self-sufficient in a region and others are virtually integrated with other clusters. These two types of supply networks demonstrate that inter-firm collaborations have emerged as a new type of manufacturing system.

Combining both developments, as Fig. 1 illustrates, a new type of manufacturing network can be derived with the characteristics of international and inter-firm relationships. The new combination provides a new operational environment for a manufacturing system to access, optimise, and operate its strategic resources. The global manufacturing virtual network (GMVN) was suggested to explore the new generation of manufacturing architecture (Li et al. 2000; Shi and Gregory 2002). Many other researches on global outsourcing and partnership also seek to develop the system with similar architecture and strategic capabilities pursuing higher value and innovation (Normann and Ranfrez 1993; Parolini 1999; Bovel and Martha 2000).

Why does a manufacturing system have to evolve into such complex relationships? Manufacturers have no other choice. In many circumstances, when there are traditional product-based competitive advantages, which even achieve the order-winning criteria, this still cannot satisfy new corporate demands. For example, a case study demonstrates that a very successful order-winning UK aerospace engine company was deeply shocked by comparing its profitability with its American competitors and by the huge pressures from its shareholders. The reasons for such unpredicted shocks were not only the cruel results of their product

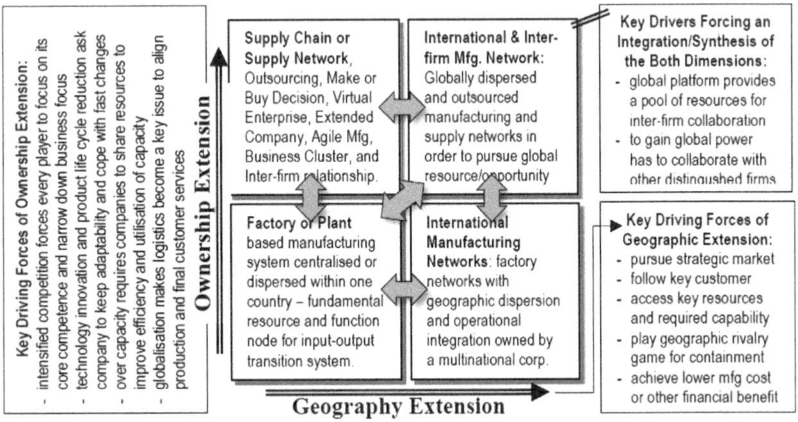

Fig. 1 Manufacturing system evolution matrix and key drivers. Adapted from Rudberg and Olhager (2003), and Shi (2004)

success and at the same time their financial failure, but also a huge challenge to their traditional ideas relating to the excellence of their product design, technology, engineering, and production. The company indeed had been very successful to gain much more orders and a greater market share than any other competitors in the world because of its modularised engine platform, advanced manufacturing technologies, and the best engine performance. However, at the same time, while the company enjoyed the advantage of more orders and production activities, its competitors have initiated to change the rules of the game. The UK company's proud engines were recognised as a commodity and integrated under its competitors' solution packages. The competitor companies were fully engaged into a new service business by providing the total solution of power to airline companies. The competitors had also outsourced a large portion of their manufacturing to their Far East suppliers, and redefined manufacturing system as a value creation system rather than product-oriented production or transition system. This was the time when the UK company realised that the traditional best manufacturing capabilities were no longer good enough for future competition and value creation. They realised that the manufacturing system and its strategy would have to be extended to match the changed rules of market competition. Therefore, the manufacturing system has to be capable of not only providing competitive products but also finding a proper position in an innovative solution to final customers.

From this case study and an understanding of the trends of the manufacturing system towards a network relationship (Fig. 1), the following lessons could be learned: (1) the traditional manufacturing strategy focussing on a product and its effective factory might not be enough, especially for creating higher business value; (2) manufacturing system has been extended into new operational space—international and inter-firm relationship—mainly because of new competition game and strategy; and (3) manufacturing system boundary changes imply manufacturing strategy also needs to be changed in terms of its contents and process.

2 Manufacturing and Its Systems

The term "manufacturing system" might be firstly introduced by R. Owen, a British utopian socialist in 1815, to mean "factory system". Since then, the term manufacturing system has been typically used to represent following two related concepts. The first one is a system approach to manufacturing emphasising dynamics and optimised action —"Manufacturing systems approaches seek to optimise the initial design to commercial product time, the design lead time and factory door-to-door time, the manufacturing lead time by considering the whole factory as a system and simplifying and optimising the performance of this complete system" (Williams 1988).

The second meaning is focused on the boundary of manufacturing systems, stressing that the manufacturing system is a unified assemblage of hardware including workers, production facilities, material-handling equipment, and other supplementary devices. Focusing more on its dynamic aspect, the manufacturing system can be defined as the conversion process of the factors of production, particularly converting raw materials into finished goods with the aim to maximised productivity.

It had been long time for many people to believe the mission of manufacturing system is to maximise productivity or efficiency, mainly because of the shortages of goods and manufacturing capacity in society. Although Skinner's paper (1969) alarmed people to pay more attention to manufacturing effectiveness or manufacturing strategic aim/mission, the time when manufacturing became a real strategic concern in product or business was the mid-1980s (Hayes and Wheelwright 1984; Hill 1993). Afterwards factory-based manufacturing system, its effectiveness, and the strategy for product's competitive advantages have regained much more attentions in both practitioner and academia worlds.

Figure 2 demonstrates a process of the manufacturing strategy ensuring a factory to provide its products' competitive advantages in market. The basic assumption of the strategy is that a company will create and capture value if its products can win its order in competition. Based on this rule, manufacturing managers design an effective system to

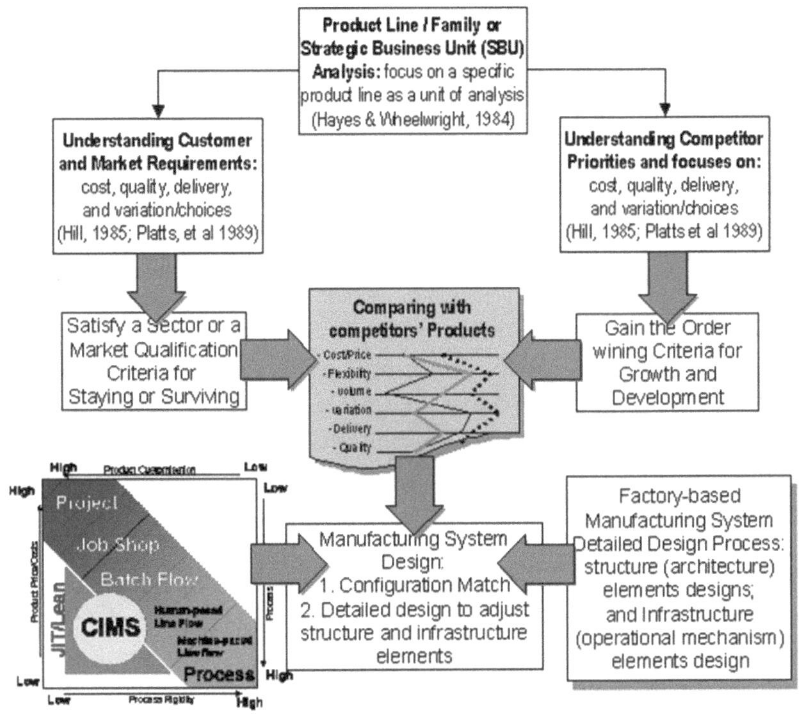

Fig. 2 Manufacturing strategy process for a product family

win orders and beat competitor, which transforms manufacturing missions from productivity or efficiency towards the competitive advantages in terms of lower cost, faster delivery, better quality and more flexible for customer choices. This fundamentally transforms manufacturing management, especially manufacturing system design.

3 Manufacturing Internationalisation Towards Geographically Dispersed Networks

Although international operations can be very different from domestic ones, it is not clear whether an internationalised manufacturing system will have any impact on manufacturing strategy. Different researchers have quite different solutions to deal with manufacturing globalisation. Sprague (1990) suggests a global manufacturing decision grid by implementing Skinner's ideas globally. The consequence might be that each dispersed factory should follow its local market demands, and therefore, there is no commonality between factories if the targeted market or customer is different. On the other hand, MacCormack et al. (1994) developed another framework for global manufacturing site locations. As high technology and regional trade blocs are moving international manufacturing towards a smaller scale and decentralisation to serve local markets with higher flexibility, not only does site location become a major issue of manufacturing internationalisation but also the manufacturing system converges towards a globalised identical system with computerised flexibility. Both solutions hold their grounds but lead to different directions. They indicate that there is not yet a comprehensive picture of manufacturing globalisation, and a greater understanding about the complexity is required.

Since the early 1980s, many academics have begun to analyse in detail the globalised manufacturing network. Hout et al. (1983) studied several global competitors' strategies, and found that the key factor in their success was to rationalise their product lines and their manufacturing and distribution systems in order to gain economies of scope from the global scale. The strategic positioning of their manufacturing facilities is very critical for global players, so they can access the markets and achieve global efficiency. Kogut (1985) and Porter (1986) built this idea into the value-adding chain concept, and suggested that the global companies needed to disperse the value-adding chain geographically to access the most appropriate resources and, at the same time, to achieve economies of scale.

Other interesting research findings include the global competition game played by two rivals (Hamel and Prahalad 1985; Yip 1992). The authors suggested that a better way to protect one company's home market is not only by defending yourself domestically but also through an attack on your rival's home market, which will not only release the pressure on own home market but also enable you to penetrate into new foreign markets. The global platform therefore provides much wider operations space to compete against competitors, harvest new markets, organise resource, and enhance capability.

Flaherty's work (1986) was very valuable with the development of another type of global coordination mechanism that goes beyond the geographically dispersed value chain. Although this coordination mechanism has been adopted by the service industry for a long time, establishing McDonald's, Kentucky Fried Chicken, Pizzas Hut and so on, her detailed studies on micro-electronics industry also analysed the globally coordinated mechanism in manufacturing operations, and provided a sound basis for the observation and exploration of international manufacturing systems. Based on her empirical research, she suggested that a network with particular patterns of geographic dispersed facilities and a shared common infrastructure and mechanism could lead to a synergy advantage in the network.

Ferdows observed the international manufacturing networks from another perspective and found that factories had different strategic roles, e.g. off-shore, source, server, contributor, outpost and lead, in the network (1989, 1997). Ferdows attempted to link strategic motivations to the role and capabilities of each factory. The problem of this model perhaps is that it overemphasises the strategic role of separate factories compared with the holistic characteristics and strategic capabilities of the network, and complexity of the roles from the strategy aspect.

Cambridge Manufacturing Group's work focuses on the identification of different network configurations on the platform of manufacturing dispersion and coordination (Shi and Gregory 1994, 1998). Eight manufacturing configurations were identified through extensive case studies, and the relationship between network configurations and strategic capabilities was explored in order to explain the current transformation towards more globally integrated or coordinated configurations. The research provides a wider scope of the manufacturing network system on the global platform and maps an evolutionary route of manufacturing system towards globalised networks.

Figure 3a illustrates two dimensions—geographic dispersion and coordinated interdependence—to map typical international manufacturing network configurations. Based on case studies in more than twenty MNCs worldwide, a new vision of manufacturing system can be transformed from the traditional factory-based single-site model (Hayes and Wheelwright 1984) towards international manufacturing networks. To understand the patterns of international manufacturing networks, a "configuration map" and seven typical network configurations are identified. These configurations provide a structured view of

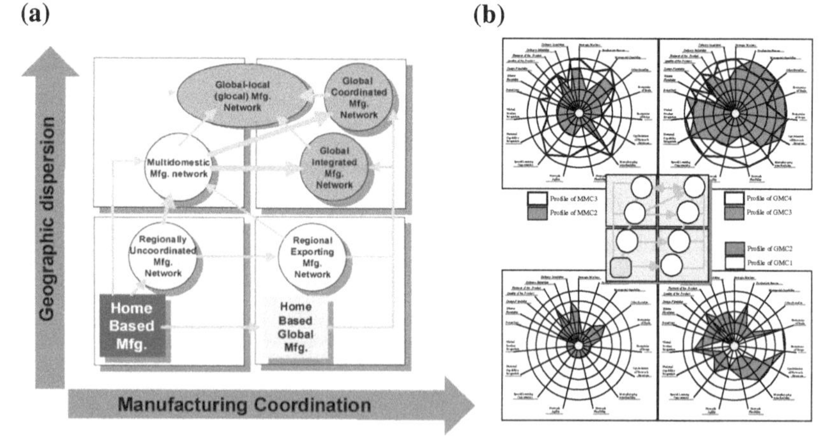

Fig. 3 International manufacturing network configurations and capabilities. **a** Int'l Mfg. Network Configuration Map; **b** Int'l Mfg. Network Capability Profile. *Source* Adapted from Shi and Gregory (1998)

international manufacturing networks and their evolutionary paths. The configuration map is compatible with established research but provides much more depth in manufacturing.

At the same time, five strategic capabilities of networks have also been identified in order to understand new mission of manufacturing network:

- Strategic resource accessibility: the ability of the network to capture some required manufacturing resources;
- Thriftiness ability: the ability of the network to support higher efficiency;
- Manufacturing mobility: the ability of the network to deploy and reconfigure resources swiftly;
- Learning ability: the ability of the network to capture and disseminate internally generated knowledge;
- Supportiveness of manufacturing network to its individual factories and products in terms of cost, quality, delivery, and flexibility.

There is a strong relationship between network configurations and the strategic capabilities as Fig. 3b demonstrates. The capability profiles of each typical configuration not only clarify that different network configurations can generate or reserve different capabilities but also provide a set of practical tools for network assessment, auditing, and design.

In summary, it is obvious that not only the internationalised manufacturing networks are different from the factory-based manufacturing system but also their missions are changed when the boundary of the system stretches. The aims of international manufacturing network can include the traditional manufacturing system's effective and efficient missions and can also reach beyond them. Its aims and compatible capabilities are driven by market and resource proximity, capability allocation and mobility, pre-emption and quick response to competition, and international diversity integration and learning. From these perspectives, the international manufacturing system has its own missions, context, contents, and therefore, its own design process which implies a new compatible manufacturing strategy process is necessary.

4 Manufacturing Externalisation Towards Inter-Firm Collaborated Supply Chains

If manufacturing internationalisation used to be recognised as a process of the internalisation of business activities (Rugman 1980), it is more often to observe a clear trend of externalisation from traditional vertical integrated firms in almost every sector. The externalisation or outsourcing makes supply chain/network subject even more popular. There are three streams of supply network studies roughly representing major characteristics in the area. All of them raise the same serious challenge to the traditional concept of the firm, especially in the field of ownership and control of resources. At the same time, the new conceptualisation of the supply network (Harland et al. 1999) also has a strong impact on the manufacturing strategy theory.

The main contents of supply network management are not new at least to production/operations management. The first stream can be traced back to the traditional inventory models and control mechanisms which used to streamline production flows between production stations or workshops. Forrester (1961) explored the bull whip effect between firms which widened the scope of studies in industrial dynamics. After developments following Porter's value chain (1985) and Japanese JIT system (Ohno 1988), especially the concept of lean production (Womack et al. 1990), inter-firm integration along the supply chain or value chain gradually emerged. Although there are many researches still focusing on restructuring and streamlining a supply network, fundamentally they all can be traced back to the classical manufacturing strategy and operations management root that is seeking product's competitive advantage or even just efficient smoothness. Related logistics and supplier development and even whole supply network architecture design can still belong to the classical framework of manufacturing strategy (Slack and Lewis 2002).

But the second stream of supply network research is quite different, it comes from not only strategic management about collaboration and value creation but also recent business practices like outsourcing and focuses on core competence, especially with respect to the recent developments in the electronics industry which include the separation between original equipment manufacturers (OEMs) and their contractual electronics manufacturers (CEMs) or electronics manufacturing service (EMS) providers. This new development goes beyond the traditional make-or-buy decisions, and it creates a new player and even a new industry that has specialised competences and innovative collaboration potentials. The trends of specialisation and collaboration between firms are not limited to the electronics industry; Ford that used to run the most comprehensive internally integrated supply chain (Ford and Crowther 1922) also separates its large portion of manufacturing forming Visteon to provide professional manufacturing services to all OEMs even including GM. This type of industry dynamics—outsourcing, specialisation and collaboration—creates another type of supply network which is no longer just for the old product family but is also for new value proposition and new strategic position in the supply or value network (Bovel and Martha 2000).

The third stream comes from clustering studies (Piroe and Sabel 1984; Keeble 1998; Porter 1998; Teece et al. 2001). This does not directly link with supply network management, but it does expose many critical characteristics of the networked system, for example the cluster's contribution to innovation and competition for vertically integrated companies. The clustering phenomenon provides a demonstration to the firm of a complementary model that is another type of organisation bridging market demands and firm's resources. It also validates the supply network with its own life power, especially in dramatically changed and innovative environments.

There are many overlaps between these three streams. However, it is clear that the boundary of supply network is blurring from a focal firm-based rigid network with integrating up and down streams tier suppliers towards virtual enterprise and industrial cluster, which has increasing dynamics, as well as an equally important position for each player that is specialised on its core competences, and also a loose collaboration relationship in a network.

Like the international manufacturing network, the externalised inter-firm supply network has extended its boundaries from the factory-based manufacturing system along the ownership dimension

(Fig. 1). This extension making the supply network a new unit of analysis also has more features than the classical factory. While the focal firm-based supply network can still be designed by following the classic manufacturing strategy process, this is not the case with the virtual enterprise, value network, and clustering because it is beyond its coverage as well as beyond the design and planning schools of strategy (Mintzberg et al. 1997). The new supply network, in contrast to the international manufacturing network, also has new missions, architectures, mechanisms, and a strategy formation process. Apart from providing support to product competitive advantages, the supply network strategy (Lamming et al. 2000) appears to strategically seek more opportunities from external markets, other companies' resources and capabilities, heading towards internal innovation, growth and higher value creation.

5 Manufacturing Internationalisation: Forming a New Decision Space

The previous two sections explain two-dimensional expansions of manufacturing systems—geographic dispersions and inter-firm collaborations. And, in the real world, manufacturing internationalisation and externalisation never work alone but always interdependently interact. The following diagram illustrates a new type of international manufacturing system emergence that is international and inter-firm collaborative manufacturing network. From a domestic manufacturer towards a global supply network coordinator, there are many paths to achieve the strategic position, by either off-shoring own resources and then localising supply chain or outsourcing parts of vertically integrated value chain and then expanding own footprints in foreign countries.

As manufacturing can be also extended as a value creations process from R&D, Design, and Sourcing to Production, Route to Market, and Servitisation (Fig. 4), its combination with the above two-dimensional model creates a new three-dimensional space as the following diagram illustrates for global value network management. The management team needs to make strategic positioning and coordination decisions within the new decision-making space.

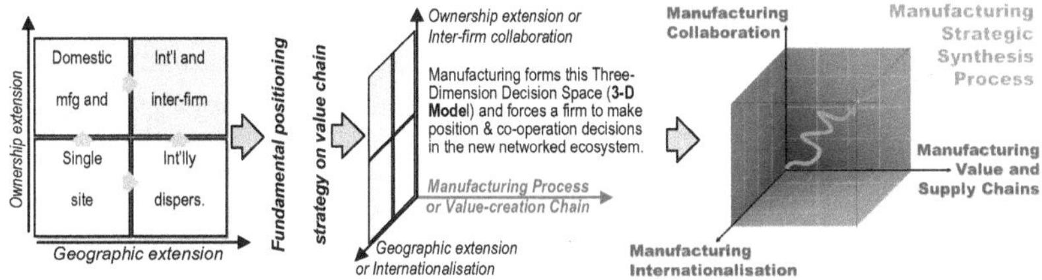

Fig. 4 An evolutionary process of manufacturing systems towards the Three-Dimensional Decision Space (the 3-D Model)

Cambridge IfM published the book—Capturing Value from Global Networks (IfM 2014)—that seeks to further visualise and explain the above three-dimensional decision model as the following diagram (Fig. 5). Figure 5 simplifies three-dimensional relationships into a 2-D plain demonstrating (1) the horizontal value creation process from R&D and production towards service, (2) the vertical dimension representing geographic dispersion, and (3) the two-dimensional plain representing the complex inter-firm collaborative relationships that are vertically cross-regions and within a similar strategic function as well as horizontally to form the value chain collaboratively. In recent academic research work as well as in this book, many scholars have dedicated their researches based on strategic business functions, such as globalised R&D networks, production and operations networks, supply and logistic networks, and after sales service networks. The globalisation generates a

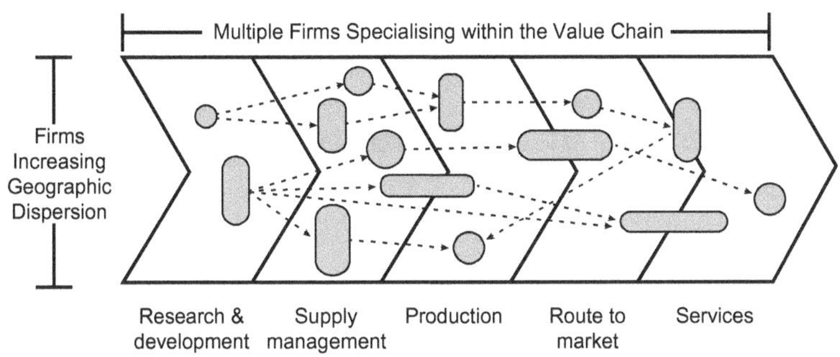

Fig. 5 Capture Value from Global Networks. *Source* IfM 2014, reprinted with permission

platform enabling not only the strategic business functions to evolve into complex network systems but also more interactive mechanisms between the functions in order to streamline the value creation process cross the functions.

Manufacturing globalisation not only creates complex global value networks for multinational corporations (MNCs) but also involves more developing nations into the globalisation. The most impressive example for such kind of involvement and development must be the Chinese manufacturing industrial growth. In the last 35 years, China emerges from a very humble position towards the largest manufacturing nation in the world, mainly through engaging its industrial capabilities with the global economy and continuously upgrading its manufacturing systems and ecosystems.

Behind the Chinese successful stories and the globalised manufacturing networks visualised in Fig. 5, there is an almost invisible network emerging during the manufacturing globalisation and developing nations' industrialisation. This new type of networks has been entitled as the global engineering networks (GENs) (Zhang et al. 2007, 2016). GENs are not easily recognisable because they don't belong to the normal business functions, such as R&D, or operations, or supply chain management, but a deeply embedded system within the existing business functions. However, GENs become so critical for both established MNCs and emerging countries' manufacturers that they are empowering strategic catching-up, organisational learning and various innovations, and nurturing a type of strategic business capability through the interactions between engineering and manufacturing.

6 Emergences of International Engineering (Towards GEN)

Contrasting to manufacturing, engineering has much longer history in human civilisations. The term "engineering" originated from its Latin root *ingeniōsus*, meaning ingenious or skilled and characterised by cleverness or originality of invention, production, or construction (Oxford Dictionary 2011). The engineering definitions have evolved with time, as shown in

Table 1 Evolutions of the Engineering Concepts

Da Vinci (the fifteenth century)	Engineering is the synthesizing art and technology by embodying the qualities of inquiry, imagination, scientific and technological rigour, vision, and creativity —quoted from Hawley (p. 16)
Tredgold (1828)	Engineering is the art of directing the great sources of power in nature for the use and convenience of human —quoted from Kirby et al. 28 (p. 2)
Kirby et al. (1956)	Engineering is the art of the practical application of scientific and empirical knowledge to the design and production or accomplishment of various sorts of constructive projects, machines, and materials of use or value to man
Rogers (1983)	Engineering refers to the practice of organising the design and construction of any artifice which transforms the physical world around us to meet some recognised need
Florman (1996)	Engineering is the art or science of making practical application of the knowledge of pure sciences
Jordan et al. (2002)	Engineering is the application of science and human experience to solve problems faced by people, which is often done in poorly understood or uncertain situations, using the available resources
Koen (2003)	Engineering methods are the strategy for causing the best change in a poorly understood or uncertain situation within the available resources and the use of heuristics
ABET (2004)	Engineering is the profession in which a knowledge of the mathematical and natural sciences, gained by study, experience, and practice is applied with judgment to develop ways to utilise, economically, the materials and forces of nature for the benefit of mankind
Dictionary.com (2004)	Engineering is the application of the scientific and mathematical principles to practical ends, e.g. design, manufacturing, and operation of efficient and economical structures, machines, processes, and systems. It means the application of science, accomplished by applying the knowledge, mathematic, and practical experience to the design of useful objects or processes, to the needs of humanity
Encyclopaedia Britannica (2007)	[Engineering is] the application of science to the optimum conversion of the resources of nature to the uses of humankind [.] the creative application of scientific principles to design or develop structures,

(continued)

Table 1 (continued)

	machines, apparatus, or manufacturing processes, or work utilising them single or in combination; or to construct or operate the same with full cognizance of their design; or to forecast their behaviour under specific operating conditions; all as respects and intended function, economics of operation and safety to life and property
Merriam-Webster (2007)	[Engineering is] the activities or function of an engineer, i.e. the application of science and mathematics by which the properties of matter and the sources of energy in nature are made useful to people, or the design and manufacture of complex products
Wikipedia (2011)	Engineering is the discipline, art, skill and profession of acquiring and applying scientific, mathematical, economic, social, and practical knowledge, in order to design and build structures, machines, devices, systems, materials, and processes

Table 1 (Zhang et al. 2014). The term is now considered broadly as an ability of directing the great sources of power in nature for the use and convenience of human (Kirby et al. 1956; Robertson 1960) and or specifically as activities of finding solutions within constraints of resources, technologies, and environments (Finch 1951; Morgan et al. 1998).

From business function perspectives, because of its design and problem solving, engineering is also usually recognised to be positioned between research and manufacturing (production). Table 2 provides a comparison and highlights the unique characteristics of engineering distinguished from research and production (manufacturing) (Zhang et al. 2014).

Based on the engineering's characteristics from the comparisons, the engineering value chain (EVC) model can be summarised as shown in Fig. 6. It identifies five critical stages as the core function and process of engineering including Idea Generation and Selection, Design and Development, Production and Delivery, Service and Support, and Recycling and Disposal (Zhang and Gregory 2011). Primary value creation mechanisms have different focuses on priority areas such as generating novel ideas, efficient product development, reliable production delivery, and flexible service offerings, along the engineering value chain.

Table 2 The Key differences between Research, Engineering, and Manufacturing

	Basic research	Engineering	Manufacturing
Philosophical stance	Concerning the truth of how the world works, the "light" question*	Concerning the underlying truth of the world as a means to the practical use, the "fruit" question and the underlying "light" question	Concerning the practical use of a product as well as the production processes, the "fruit" question
Main output	Theories, assumptions, or hypothesises	Design, solutions, or prototypes	Products, artefacts, or commodities
Principal method	Rational reflection or searching for empirical evidence, relying on the knowledge of individual scientists	Interaction with systems through attempts to build, fix, or modify them, relying on the know-how of engineers and organisations	Implementation of instructions or procedures in a reliable and efficient way, relying on the expertise of organisations
Drivers for task choice	Random problems driven by the curiosity or desire of a scientist	One-off problems given by the customer, colleagues, or government	Repetitive tasks co-defined by the customer and the manufacturer

**Note* 1 "Fruit" and "light" are the two main aims of classic sciences according to Francis Bacon (RAEng 2010). "Fruit" stands for the application of science to enable to anticipate and to control nature. "Light" stands for understanding how the world works

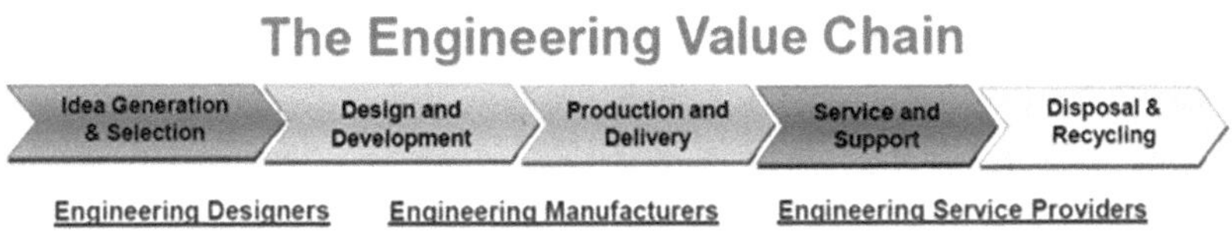

Fig. 6 Engineering value chain model. Adapted from Zhang and Gregory (2011)

Although it is somehow similar with the IfM manufacturing model (Race to the Top), the key difference is the repeatability—manufacturing, particularly the production part, tends to be more mass production oriented, but engineering more focused on bespoken issues or projects. There are some strong overlaps in the later stages where engineering plays a strong role to solve various types of problems and generates all kinds of solutions, while the typical manufacturing more focuses on routine operations to ensure system reliability and output conformance and consistence.

During business globalisation, particularly the foreign direct investments (FDI) push the whole business value chain to international, engineering as a critical function has been evolving into Global Engineering Networks (GENs) (Zhang et al. 2007). The GENs not only fundamentally changed multinational corporations (MNCs) engineering organisation but also strategically assisted many developing nations' businesses engaging with the global resources and capabilities. GENs emerge as one of the critical competitive weapons in global competition and capability development.

Zhang et al. (2008) illustrate an interesting evolutionary journey of engineering as a function towards the GENs. From the very original craftsman skills of the human beings at the top level, the engineering function and organisation gradually adapt themselves reactively or proactively towards more globally dispersed and interdependent networks. As the global production networks (GPNs) tend to consolidate themselves into a few coherent configurations (Fig. 3), research also finds that the GENs organise themselves into four types of configurations with different unique strategic orientations for effectiveness from both internal coordination and external function perspectives.

In order to exploit knowledge about GEN and help engineering companies to achieve better global engineering network redesign, the Cambridge IfM team has also developed a process-based GEN design framework developed in Zhang et al. (2007). It highlights four phases from understanding the GEN contexts, auditing the capability, scrutinising the configuration, and searching for excellences through continuous improvement during the engineering operations.

In summary, manufacturing business globalisation not only creates globalised production networks (GPNs) owned by multinational corporations and globalised supply networks (GSN) formed by inter-firm collaborations but also changes engineering functions and reconfigures its organisations towards the GEN. Table 3 highlights their distinguished characteristics and relationships (Zhang and Gregory 2011).

Table 3 A Comparison between GPN, GEN, and GSN

Configuration Elements	International manufacturing networks (Shi and Gregory 1998)	Global engineering networks (Zhang et al. 2007)	International supply networks (Sari and Gregory 2008)
Network structure	The structure of international manufacturing networks should consider individual plant's characteristics and the degree of geographic dispersion	Critical aspects to describe network structures include: geographic dispersion, resources and roles of engineering centres, and rationales for network structure design	Network structure refers to the various operations within the supply network and their integrating mechanisms
Operations processes	Key operations processes of international manufacturing networks include operational mechanisms, dynamic response mechanisms, product lifecycle management, and knowledge transfer processes	Operations processes describe the flows of information and materials among network members, e.g. new product development, safety management, procurement, etc.	Operations processes refer to the flows of materials and information between and within key unit operations
Governance system	Governance system of international manufacturing networks should include	Key aspects of network governance include the authority	Network governance refers to the role of and coordination mechanisms

(continued)

Table 3 (continued)

Configuration Elements	International manufacturing networks (Shi and Gregory 1998)	Global engineering networks (Zhang et al. 2007)	International supply networks (Sari and Gregory 2008)
	horizontal/vertical coordination, dynamic capability building, and network evolution	structure and the performance measures	between key network partners
Support infrastructure	Not emphasised in the above research or combined with the other elements	Support infrastructure includes engineering tools and IT systems	Not emphasised in the above research or combined with the other elements
External relationships	Not emphasised in the above research or combined with the other elements	Network configuration should consider relationships between network members, e.g. suppliers, customers, or users	Relationships refer the roles and interactions between key network partners

7 Conclusion: Continuous Adaptations of Industrial Systems

Based on Introduction of this chapter and the research into the evolutions of industrial systems in the last 30 years, the new system boundaries for not only industrial systems including all types of intra-firm and inter-firm networks but also the business ecosystems can be recognised in order to cope with the uncertainties of both product and manufacturing systems. A new generation of "product–process matrix" (Hayes and Wheelwright 1975) therefore can be proposed as Fig. 7 demonstrates to meet the new requirements. The business ecosystem as a solution to cope with new challenges is discussed to call for more research and understanding in production and operations management (P/OM)

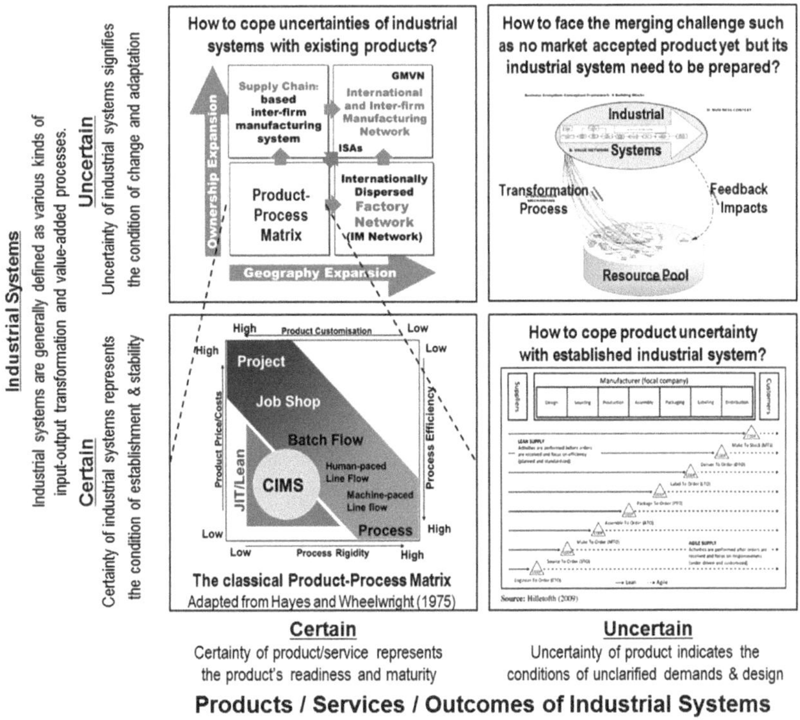

Fig. 7 New product-process matrix for environmental uncertainties

community. The new matrix explains the role of operations ecosystems in the POM body of knowledge.

The new product–process matrix, operations ecosystems do have their clear boundaries distinguishing from the classical industrial systems or value chains. At the top-right corner, it illustrates dynamic characteristics of the engineering/manufacturing value chains. Both diagrams conceptually offer a new definition and a different vision about operations ecosystems—an operations ecosystem is an interconnected and interactive community and continuously nurtures new industrial systems to create business value.

References

Accreditation Board for Engineering and Technology (ABET). (2004). *Criteria for accrediting engineering programs.* New York: Accreditation Board for Engineering and Technology.

Bovel, D. & Martha, J. (2000). From supply chain to value net. *Journal of Business Strategy,* July/August.

Brewer, A. M. et al. (eds.). (2001). *Handbook of logistics and supply-chain management.* Amsterdam: Pergamon.

Dictionary.com. *Engineering.* Retrieved December 23, 2004, from http://www.dictionary.com/.

Encyclopaedia Britannica. *Engineering.* Retrieved February 12, 2007, from http://www.britannica.com/eb/article-9105842.

Finch, K. J. (1951). *Engineering and western civilization.* London: McGraw-Hill.

Ferdows, K. (1997). *Making the most of foreign factories.* Harvard Business Review, March-April.

Flaherty, M. T. (1986). Coordinating international manufacturing and technology. In M. E. Porter (ed.), *Competition in global industries.* Harvard: Harvard Business School Press.

Florman, S. C. (1996). *The existential pleasures of engineering.* New York: St. Martin's Griffin.

Ford, H. & Crowther, S. (1922). *My life and work.* Messina.

Forrester, Jay W. (1961). *Industrial dynamics.* The MIT Press.

Hamel, G., & Prahalad, C. K. (1985). *Do you really have a global strategy?, 63*(4), Harvard Business Review, Jul/Aug 85.

Harland, C. M., Lamming, R. C., & Cousins, P. D. (1999). Developing the concept of supply strategy. *International Journal of Operations and Production Management, 19*(7).

Hayes, R. H., & Wheelwright. S. C. (1975). *Link manufacturing process and product life cycle.* Harvard Business Review, Jan-Feb.

Hayes, R. H., & Wheelwright. S. C. (1984). *Restoring our competitive edge: Competing through manufacturing.* John Wiley & Sons.

Hill, T. (1993). *Manufacturing strategy: The strategic management of the manufacturing function* (1st ed. 1985, 2nd ed.). London: Macmillan.

Hout, T. M., Porter M. E., & Rudden, E. (1983). How global companies win out. In R. G. Hanermesh (ed.) *Strategy Management*, John Willey and Sons, Inc.

IfM (Institute for Manufacturing). (2014). *Capturing value from global networks*. www.ifm.eng.cam.ac.uk/insights/global-supply-chains/capturing-value-from-global-networks/.

Jordan, W. B., Elmore, S., & Napper, A. (2002, September). A biblical perspective on engineering ethics. In *Proceedings of the national faculty leadership conference, LA*.

Keeble, D., & Wilkinson, F. (1998). Collective learning and knowledge development in the evolution of regional clusters of high technology SMEs in Europe. *Regional Studies*, 33(4), 295–303.

Kirby, R. S., Withington, S., Darling, A. B., et al. (1956). *Engineering in history*. New York: McGraw-Hill.

Koen, B. V. (2003). *Discussion of the method—Conducting the engineer's approach to problem solving*. New York: Oxford University Press.

Kogut, B., (1985). Designing global strategies: Comparative and competitive value-adding chains. *Sloan Management Review, Summer*.

Lambert, D. M., Cooper, M. C., & Pagh, J. D. (1998). Supply chain management: Implementation issues and research opportunities. *International Journal of Logistcs Management, 9*(2), 1–19.

Lamming, R. et al. (2000). An initial classification of supply networks. *International Journal of Operations and Production Management, 20*.

Li, X., Shi, Y., & Gregory, M. (2000). Global manufacturing virtual network. *EurOMA, 2000 Conference*.

MacCormack, A. D., Newman, L. J., & Rosenfield, D. B. (1994). The new dynamics of global manufacturing site location. *Sloan Management Review, Summer*.

Merriam-Webster. (2007). *Engineering*. Retrieved February 12, 2007, from http://www.merriamwebster.com/dictionary/engineering.

Mintzberg, H., et al. (1997). Strategy safari. Free Press.

Morgan, R. P., Reid, P. P., & Wulf, W. A. (1998). The changing nature of engineering. *ASEE Prism, 7*, 13–17.

Normann, R., & Ramirez, R. (1993, July-August). From value chain to value constellation: Designing interactive strategy. *Harvard Business Review*, 65–77.

Ohno, T. (1988). *The toyota production system* (trans). Cambridge, MA: Productivity Press. 1978.

Oxford Dictionaries. (2011). *Engineering*. Retrieved December 30, 2011, from http://oxforddictionaries.com/definition/engineering?q=engineering.

Parolini, C. (1999). *The value net: A tool for competitive strategy*. John Wiley Books.

Piroe, & Sabel. (1984). *The second industrial divide: Possibilities for prosperity*. New York: Basic Books.
Porter, M. E. (1985). *Competitive advantage* (pp 11–15). New York: The Free Press.
Porter, M. (1986). *Competition in global industries*. Harvard Business School Press.
Porter, M. (1998). Clusters and the new economics of competitiveness. *Harvard Business Review,* December.
Robertson, S. A. (1960). *Engineering management*. Glasgow: Blackie & Son.
RAEng. (2010). *Philosophy of Engineering*. London: The Royal Academy of Engineering.
Rogers, G. F. C. (1983). *The nature of engineering*. Helsinki: The Macmillan Press.
Rudberg, M., & Olhager, J. (2003). Manufacturing networks and supply chains: An operations strategy perspective. *Omega, 31,* 29–39.
Rugman, A. (1980). A new theory of the multinational enterprise: Internationalisation versus internalisation. *Columbia Journal of World Business, 15*(1). Spring.
Sari, J., & Gregory, M. (2008). A supply network configuration perspective on international supply chain development. *IJOPM, 28*(5), 386–411.
Shi, Y. J., & Gregory, M. (1998). International manufacturing networks—To develop global competitive capabilities. *Journal of Operations Management, 16,* 195–214.
Shi, Y., & Gregory, M. (2002). Global manufacturing virtual network (GMVN): Its dynamic position in the spectrum of manufacturing collaborations. In U. J. Franke (ed.), *Managing virtual web organisations in the 21st century: Issues and challenges*. Idea Group Publishing.
Shi, Y. J. (2004, March). A roadmap of manufacturing system evolution: From product competitive advantage towards collaborative value creation. In Y. Chang et al. (Eds.), *Evolution of supply chain management—Symbiosis of adaptive value networks and ICT*. USA: Academic Publisher. ISBN 1-402-7812-9.
Slack, N. & Lewis, M. (2002). *Operations strategy*. Pearson Education Limited.
Sprague, L. G. (1990). Strategic analysis for global manufacturing. In P. E. Moody (ed.), *Strategic manufacturing: Dynamic new direction for the 1990s*. Business One Irwin.

Teece, et al. (2001). The geography of innovation and economic clusters. *Industrial and Corporate Change, 10*(4).
Tredgold, T. (1828). https://en.wikipedia.org/wiki/Thomas_Tredgold.
Wikipedia. (2011). *Engineering*. Retrieved December 30, 2011, from http://en.wikipedia.org/wiki/Engineering.
Williams, D. J. (1988). *Manufacturing systems: An introduction to the technology*. Chapman and Hall.
Womack, J. R., Jones, D. T., & Roos, D. (1990). *The machine that changed the world*. Rawson Associates: New York.
Yip, G. (1992). *Total global strategy: Managing for world-wide competitive advantage*. Prentice Hall.
Zhang, Y., Gregory, M., & Shi, Y. (2007). Global engineering networks (GEN): The integrating framework and key patterns. *Proceedings of the Institution of Mechanical Engineers, Part B: Journal of Engineering Manufacture, 221*, 1269–1283.
Zhang, Y., Gregory, M., & Shi, Y. (2008). Global engineering networks (GEN): Drivers, evolution, configuration, performance, and key patterns. *Journal of Manufacturing Technology Management, 19*(3), 299–314.
Zhang, Y., & Gregory, M. (2011). Managing global network operations along he engineering value chain. *IJOPM, 31*(7), 736–764.
Zhang, Y., Gregory, M., & Shi, Y. (2014). Managing global engineering network (GEN) part I: Theoretical foundation and the unique nature of engineering. *Journal of Engineering Manufacturing, 228*(2), 163–171.

Authors' Biography

Yongjiang Shi is research director of the Centre for International Manufacturing (CIM) at the Institute for Manufacturing, Cambridge University. He has undertaken extensive research into international manufacturing network configurations. His research interests include global manufacturing networks, business ecosystem development, the strategies of multinational corporations in China, and sustainable industrial systems.

Yufeng Zhang is Senior Lecturer of Operations Management at the Business School of the University of Birmingham, UK. He had an earlier industrial career with responsibilities for project management and new business development, and continues to work very closely with industry aiming to help companies cope with challenges and opportunities in managing complex global engineering networks. He has researched and published in the areas of engineering management, network operations, international manufacturing, technology management and innovation.

4

Engineering Services: Unpacking Value Exchange

Florian Urmetzer, Andy Neely and Veronica Martinez

1 Value Exchange Among Engineering Services

Many commentators discuss value in manufacturing. The aim of this chapter is to unpack the concept of value, exploring what it means in a practical setting of engineering services using the service staircase. At one level, one can consider value in terms of the product being provided—does the product meet the customer's requirements and hence create value for the customer? However, customers use products to achieve outcomes. So value can also be thought of in terms of the efficiency and

F. Urmetzer (✉) · A. Neely · V. Martinez
Institute for Manufacturing, Engineering Department,
Cambridge University, Cambridge, UK
e-mail: ftu20@cam.ac.uk

A. Neely
e-mail: adn1000@cam.ac.uk

V. Martinez
e-mail: vm338@cam.ac.uk

Y. Zhang and M. Gregory (eds.), *Value Creation through Engineering Excellence*,
https://doi.org/10.1007/978-3-319-56336-7_4

effectiveness with which the product enables customers to achieve the outcomes they want—does the machine create the most value throughout its lifetime? Another perspective on value is to consider the two-way flow. Manufacturers **create value for customers**, and customers in turn **create value for manufacturers**, by, for example, continuing to use a manufacturer's services to support the products through life. In this case the customer creates economic value for the manufacturer. Additionally by feeding back data on operations of the product during its working life, the customer allows the manufacturer to improve future generations of the product (Fig. 1).

The creation of value for the customer through services brings the customer and the firm into closer proximity. More feedback is given by the customer to the machine manufacturer throughout the lifetime of the machine, which also enables the manufacturer to understand the business model of its customer better. Furthermore, the option is created for the manufacturer to interact directly with the machine and, hence, obtain data and insights into its use. When the customer brings a machine in for service or sees the manufacturer as the place to obtain spare parts, there are automatically more exchanges between the customer and the firm. This brings several opportunities of engagement, not only for gaining an in-depth understanding of how machines are used, and for additional sales opportunities through timely information on the status of machines and the customer's market situation, but there are also options for the extension of interaction through additional services to integrate the manufacturer into the customer's business model. When thinking about the level of service provision to the customer by a machine manufacturer, there are different levels that can be referred to as the service staircase.

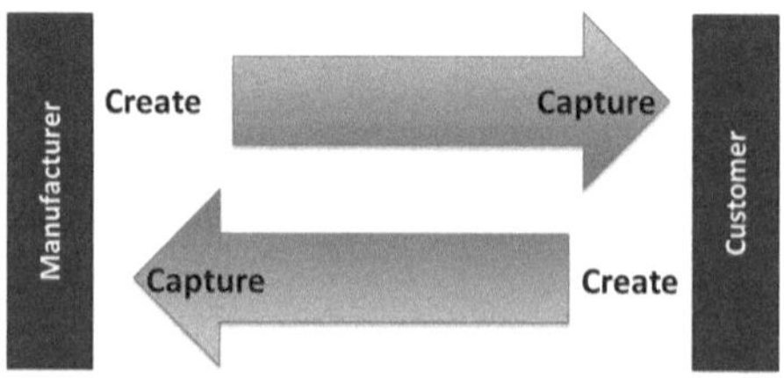

Fig. 1 The creation and capture of value

Figure 2 shows five service provision levels that firms have already implemented for their customers. The different levels of service offered to customers are usually interdependent; for example, without a working spare parts' business the offering of solutions will be problematic. This journey is often referred to as servitisation (Vandermerwe and Rada 1988). The next section will look at the value creation and capture of the manufacturer taking the structure of the service staircase. The following section will then focus on value creation and capture with a focus on the customer. Before offering some conclusions, details will be provided about asset management and how it can act as an integrator between the customer and the manufacturing company.

Even the simple examples above illustrate the complexity of defining value in manufacturing services—it is multidimensional and involves mutual exchange. In the rest of this chapter we will unpack the concept of value more fully, exploring it from the perspective of both a manufacturer and a customer. This chapter consists of two main sections. In the first section we will explore how manufacturers are able to create value through services. The second section will create an insight into how manufacturers' customers can create value through services created for them. Both parts will be structured based on the service staircase shown

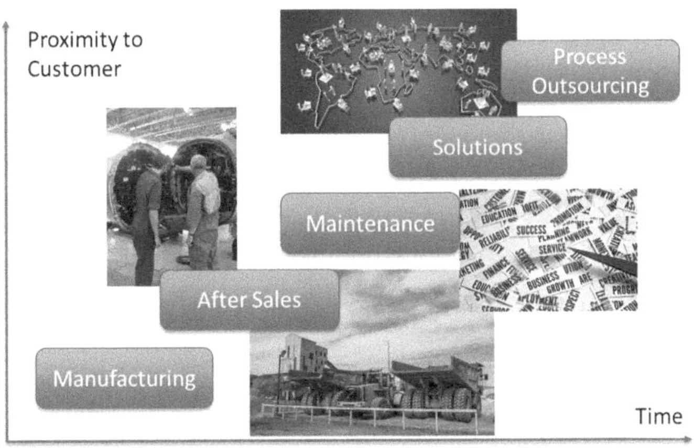

Fig. 2 The service staircase. (adapted from Turunen (2012))

in Fig. 2, in other words the offering of simple services, which increase in complexity. Finally, a summary of the value delivery will be offered.

2 The Manufacturer's View: Unpacking Value Through Services

The importance behind offering services is usually driven by the opportunity to create new revenue streams for the manufacturing firm. However, offering services also provides the opportunity to influence the design and manufacturing process through additional feedback from the customer and the gathering of information about the machines' use, as well as the customer's business model. Because the firm is moving closer to the customer and understanding the use of the machines more, there is a good opportunity to aggregate feedback and information and feed them into the design process. This is in addition to the opportunity to design new services enabling the growth of the manufacturing firm through servitisation. This section will view the value captured and created

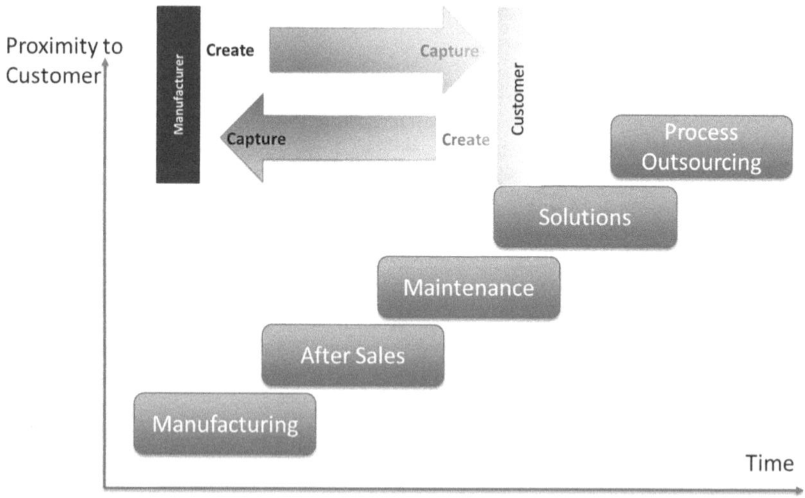

Fig. 3 The service staircase, focusing on value for the manufacturer

throughout the different parts of the service staircase, starting with spare parts' services and concluding with process outsourcing (Fig. 3).

2.1 The Importance of Spare Parts

When thinking about the second level, the sale of spare parts and what the manufacturer can create and capture in value for itself, additional revenue may be one of the main drivers. It may, however, also be that there is the intention to produce other services, which will not be implementable if there is no spare parts' distribution in place. In any case, the manufacturer will create value if a spare parts' business is in place. When interviewed, some heavy machine manufacturers indicated that there is market potential for more revenue created by selling spare parts for a machine throughout its lifetime. There will be a dependency on the type of product manufactured and its need for spare parts; however, it deserves consideration.

Having spare parts' offerings means that the manufacturing company needs to produce them and have storage, as well as distribution, channels in place. All this with the goal of allowing timely delivery of the spare part to the end-customer when necessary. In other words, when for example a machine broke down and needs a spare part to function again.

With many manufactured spare parts' products there is a grey market, where other manufacturers produce spare parts under their own brand name for products manufactured by others. This can be seen as value slippage, as there is a lost sales opportunity for the manufacturer. However, there are other risks associated with grey-market spare parts. If the quality of the spare part produced by a third party is lower than required, there is a potential impact on the machine's lifetime or other knock-on effects on the uptime of the machine.

Cases were observed whereby the manufacturer purchased a grey-market spare parts' supplier to buy in the capability and the companies' market share in the spare parts' market. In the case described, the manufacturer even kept the acquired supplier under its existing brand name to keep it separate. This allowed instant access to the existing spare parts' market and its value creation, as well as the existing distribution channels. However, it also allowed tight control to be maintained over the spare parts' quality of a large share of the market, thus also protecting the manufacturers' credibility. The above example shows that there are manufacturers that create value for their firm from spare parts.

2.2 Maintaining the Customer's Assets

Through the offering of maintenance the supplier has the option to give further support to the customer. This is the classic repair and warranty fulfilment approach of, for example, a car garage. Hence, the starting point is that the retailer, the dealer and the focal firm have the option to receive spare parts for the machine manufacturer through a distribution system and can offer the repair and maintenance of machines. It always represents high value for the customer to have access to the timely and efficient repair or maintenance of machines, and therefore, the process should be focused around the customer's needs (see later in this chapter). The creativity of offerings can be endless. One example is a heavy machinery manufacturer offering refactoring of a machine, whereby an old machine is delivered to the workshop and will be made not only operational again but also close to new. This means exchanging all parts that are worn out. The process is less expensive than purchasing a new machine, and hence, machine owners have an interest in the process. The acceptance of these offerings by the customer may differ depending on the economic situation. Higher uptake of such offerings is expected when there are higher economic risks, or market volatilities. So a decision to refactor as a form of maintenance may receive more uptake in economically uncertain times or with economically strained customers. Interestingly, insights have been obtained throughout the offering of refactoring maintenance. A significant effect was that even that the machine was technically similar to new, the machine operator was not particularly impressed by the performance of the machine and complained to the owner. When the basic package included the full renewal of the operator cabin, these complaints stopped. Two things can be learned from this. First, there is a need to review and enhance offerings constantly, and second, the look and feel of a machine is important.

It has recently emerged that there are ways to obtain information from a machine through, for example, sensors, but also through contact with the customer. The data received from a machine can easily be translated into information, helping to implement advanced asset management systems including, for example, the prediction of failure, or more. This can only be achieved when sensors have been included in the design and production of the machine. All of the above are ways in which the manufacturer can capture value through maintenance.

2.3 Offering Solutions as a Manufacturer

The next stage in the service staircase is to offer solutions for customers. A solution has in the past also been described as a full service contract. The idea is that "a comprehensive bundle of products and/or services fully satisfies the needs and wants of a customer related to a specific event or problem" (Stremersch et al. 2001). In general, a higher intensity in service contracts for manufacturers is evident in studies. This is based on maintaining competitiveness in the advent of commoditisation, declining growth and falling profitability in core product markets. Related services are sold through dedicated service divisions, which are designed to take advantage of the commercial opportunities of servicing an installed equipment base rather than just one machine. Additionally, the implementation of integrated solutions offerings is used to increase the competitiveness of core product offerings under industry conditions (Salonen 2011).

Examples for solutions offerings are twofold. One example is to guarantee uptime for one or more machines, and another is to adopt a more consultative approach, helping the produced machine to work better within the environment.

The first option for the manufacturer is to guarantee uptime by being paid by outcome. The process is relatively simple and can be taken to the extent that the manufacturing company takes over the risk by offering compensation for machine failure. The basis for this is that, instead of selling a machine to a customer, the outcome value of the machine is sold. This means, for example, for a gas engine producer, that instead of selling the gas engine to the customer, electrical energy and heat are sold. Electrical energy is produced by the gas engine's rotation, and heat is produced as a by-product of the engine combustion process. This can be of great interest to the customer (see next section). However, a manufacturer with good processes in place, and who understands the risks, has the opportunity to gain greater monetary value over time and a continuous flow of revenue instead of a one-off sale. In addition, as the machine maintenance and servicing are included in such a solution offering, there is a risk of lowering costs or losing the customer to other service providers and/or the customer purchasing parts from grey-market parts' providers. Hence, the value chain is covered and creates value for the manufacturer.

The second example is to help the customer to obtain more value from ownership of the machine and to offer this as a service. Indeed, there are machine manufacturers who offer this support. One simple example is to check the oil of large machines in a laboratory on a regular basis. The oil is tested to give an indication of not only lubrication—in other words whether the oil needs changing—but also metal and aluminium content, which is an indication of increased wear and tear of the engine. A more complex example is a wind turbine manufacturer who supports its customer in getting the most from its investment. It supports by continuously optimising the wind power plant, including the use of software and solutions that increase the production and profitability of the wind power plant. The best time to stop wind turbines for maintenance is either when there is no wind or when the electricity they produce is at its lowest cost. Weather prediction models and public data, as well as past data on wind and weather patterns, are used to anticipate any changes.

In conclusion, this level of the service staircase represents the highest complexity to reach without the prior level functioning. Without working maintenance and parts' businesses it is more complex for a manufacturer to offer solutions. There are parts of a solution that can be outsourced to other companies, for example, the transport of machines; however, the manufacturer can maintain a network of reliable partners who can be included for either specialist repairs or specific locations. These networks of partners are often referred to as ecosystems. The next section will discuss process outsourcing, which requires that such an ecosystem is understood and partners are used for collaboration.

2.4 Process Outsourcing to Create and Capture Value

The final level of the service staircase concerns the outsourcing of entire processes, and the types of process are endless. However, this level continues to show that offering services provides the manufacturer with the opportunity to move close to the customer's business and to create value for its operations by fully integrating itself into the customer's operations. This enables the machine manufacturer to move into a space where the outcome of a process is guaranteed, instead of simply providing part of a process. An example of such an outcome is the excavation of material in a quarry. In this case, the machine manufacturer is in charge of the process of excavation and placing the material in a stone crusher. This includes taking the material up, transporting it in the quarry and loading the crusher with the material. Therefore, the manufacturer, instead of supplying single machines or guaranteeing their uptime, supplies all the machines involved in the supply chain and monitors, optimises and manages them. Payment in this case can be structured by charging for material loaded into the crusher. These processes can be structured and optimised using extensive integration of technology. This not only increases the income of the manufacturer by providing the services of taking over the process and risk and spare parts for all the machines, but it also enables the manufacturer to test new machinery and technology. When interviewed, a machine operator stated that technology, like self-driving dump trucks, can be implemented when the process is owned by the manufacturer.

In greater detail, the value addition for the manufacturer is manifold, but the influencers need to be understood. The manufacturer takes over the risk of the operation and, hence, needs to understand this risk in detail. This includes all influencers that may have an impact on shaping the process, and the outcome is therefore different to what is expected. The market needs to be understood, including its seasons, cycles and volatilities.

The first section of this chapter argues that in order to deliver a service, the end value must be understood. This is what the customer wants to sell in the first place and also how, when and who the customer will be. For the quarry business, for example, this may mean there is the potential to increase the amount of ore in the rock that is excavated; there is a good chance that this will increase the value of the operation and, hence, the income of the manufacturer. On the contrary, if the amount of ore in the rock is not monitored, then the operation may become more costly. The logical question is therefore whether it can be predicted if the market is volatile, and the customer will need to move with the market and cover peak demands; it is also important to know and respond to this. As a machine manufacturer this could be a good time either to maintain or even to exchange machinery with other businesses with higher demands. It is evident that there are many opportunities for manufacturers created by supplying processes to the customer. Most importantly, however, this helps them to move closer to the customer, to understand the customer's business model and business better and to be embedded in the same.

The next section will look at the other side—how the customer thinks about value creation—and will use examples to provide a better understanding of the issue.

3 The Customer's View: Unpacking Value for the Company

How the company or owners obtain the most value out of the ownership of a machine is relatively simple: getting it to do as much as possible of what it was purchased for, whether it was moving earth, creating energy or moving people from one point to another. This value and how it is created or captured depends to a large extent on the business owner and its business model. One example is a building company owning a digger. For the owner of the building company the amount of earth moved does not matter as much as, for example, for a quarry. The owner of a building company is more interested that the hole in the ground created by the digger and its operator is of a specific diameter, so that it will fit the foundations of the house accurately. When interviewed, building company owners confirmed that they don't mind which brand of digger their company owns. Besides the need for the machine to be well priced and maintainable, the operator of the digger needs to be able to work well with the machine and to produce the right-sized hole in the ground for the foundations of the house. If the size of the foundations is not accurate, the building project will encounter problems. Most diggers on the market support all their needs.

When interviewing the operators, it becomes evident that their value focus shifts again and they want to work with a machine with which they can feel what they are digging. A digger operator wants to be able to say when he or she is digging into changing materials. For example, a digger operator for a building site will be able to feel that he or she is digging into a drain or a cable and should be able to stop in time to reduce the damage. Another example is a large stone in a quarry being dug into. The operator lifts the shovel from the digger to go above the stone and get the large stone with the next load (Fig. 4).

Another influencer is economic impacts from the market. If the market demand for what is produced is reduced, the value of the machine

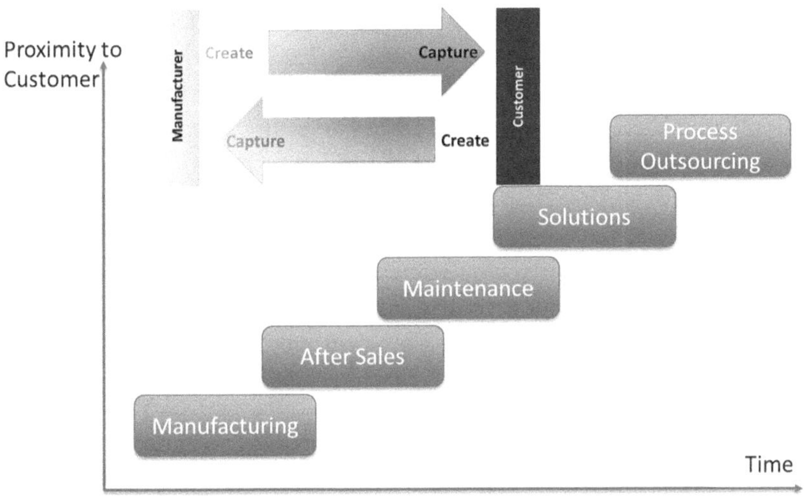

Fig. 4 The service staircase, focusing on value for the customer

is also reduced. The value propositions for the aforementioned machine are extremely diverse, depending on the actor. Therefore, the value propositions for the machine throughout its lifetime are very different. This value proposition throughout the lifetime of the machine becomes clearer when thinking about a harvester used only in harvesting season. The value proposition of ownership of the machine changes instantly when the harvesting season begins. When the machine breaks down during harvest, losses are extremely high, as the harvest will not progress.

Knowing the priorities of the use of the machine enables the supplier to create value for the machine owner. Hence, when a machine breaks down, the value proposition for the owner of the machine changes, as well as when maintenance needs to be done. The next part of this section will look at the change in value, first, of parts' sales in detail, then at maintenance and, finally, repair.

3.1 The Value of Buying Parts to the Machine Customer

Parts' availability should be seen as a baseline for successful service operation for a manufacturing company. Hence, if the parts are not available for the machine owner or the wider ecosystem within a reasonable timeframe and at a reasonable price, the value of ownership decreases. Value as a concept here can be very diverse. Taking the example of a harvesting machine during harvesting season, the acceptable timeline for the supply of a spare part needed to repair the machine will be considerably shorter than outside harvesting season. In conclusion, the customer will see parts' supply in different ways, depending on how and when the value of the machine is created.

One example comes from the European market and a heavy-equipment manufacturer. The customer in question said to its country representative that the machine would be moved over a European border for a large building contract. The dealership of the manufacturer said this would not be a problem and that they could still supply parts. However, when one of the machines failed and needed a spare part, it was not possible for the dealership to delver to the other country as it was the territory of another dealership. The customer was asked to cross the border again to pick up the part. The customer was not satisfied.

When seeing the supply of spare parts as a business model, the distribution channels and also operations can be defined very clearly. For example, the fastest delivery can be offered by an external direct courier, or the delivery route of a delivery truck can be prioritised, depending on the urgency or even importance level of the customer based on turnover.

In conclusion, a working spare parts' business can be seen as the baseline of not only a working service business but also to create value for the customer, allowing the customer to obtain more value from the machine and to capture value for the company by providing additional revenue.

3.2 Gaining Value from Maintenance

Maintenance brings additional value to the lifetime of the machine, whereby a good maintenance schedule for a machine reduces breakdowns and, hence, risk for the owner. Maintenance carried out well enables the machine manufacturer to create value for itself in the form of additional income and regular contact with the customer. This includes the manufacturer supplying its own spare parts to the customer. However, there can also be a regular conversation with the machine owner about the status of the machine, if the machine needs to be upgraded or is still the right fit for use by the customer. Both sides are open to the opportunity of value creation. Next to a repair situation, for example, the customer is in a calmer state, whereby a conversation is potentially more fruitful.

A well-structured customer relationship management system for a machine manufacturer provides the opportunity to support the customer in its decision-making and to create value for both sides. Taking the example for the harvester above, calling the customer prior to the harvesting season and offering machine maintenance will be well received. This is capturing value for the manufacturer, but also creating value for the customer, as it will reduce the risk of breakdown during the harvesting season. The result will be a more satisfied customer.

When considering management of the maintenance schedule for a machine fleet, an asset management system (AMS) is an important tool for value creation by the customer. Asset management can be defined as a tool supporting the implementation of asset management policies, for specific, measurable, achievable, realistic and time-bound asset management objectives, and clear plans to achieve those objectives (IAM 2012). The AMS will help to define not only the maintenance schedule on the basis of use but also how much value the firm is realising from ownership of the machine. The latter is also a risk-based decision. The definition of when the machine has served its purpose and should be exchanged is when the risk of breakdown of the machine and higher maintenance or repair costs become too high. In interviews, companies stated that this goal is movable, depending on the economy of the market and the financial ability of the owner company. One example was that during the financial crisis fleet owners had to make the decision that taking higher risks with assets would be better than making new purchases, meaning costs in the books. There may also be the need to decide which machines to retire earlier or to use less to reduce risk. If the economy is down and production has to be reduced, a decision must be made about machine use and retirement. If the AMS is defined in the right way, it can provide an overview of its use based on data from the machine or recorded throughout its lifetime. The information gathered about the machine will facilitate an informed decision by the management about which machine to retire or which to increase the usage of. One example of this is the stockpile management of arms. Every country has a certain amount of arms, which are kept for defence purposes. Taking a complex missile, the manufacturer guarantees the use of the arms for 10 years. After its storage lifetime it will be destroyed. The 10 years are calculated based on specific environmental conditions, meaning that a change in these conditions will have an impact on the lifetime of the missile. When it has been stored in a deep storage bunker for 10 years and not moved, its lifetime can often be extended for 5 years. However, generally the problem is the provenance of the missiles over their lifetime. Therefore, the recording of environmental data and the tracing of movements become more important. Specifically in the defence industry, this is problematic, as most of the movements are not traceable. On the other hand, the recording of data in an AMS has the potential to save large sums of money and, hence, create significant value for the customer if the lifetime of the missile can be extended.

The literature states that it is important to see AMS as a flexible system, since the practical or strategic realities of modern-day asset management need to be addressed. The issue is that requirements are changing quickly and the system therefore needs to be flexible. This is also due to the changing economic environment described above. One way to understand the flexibility of the AMS is to do scenario analysis to ascertain whether the AMS supplies the right information for the scenario and enables the right decisions to be made for the scenario. Hence, a list of test scenarios must be defined and the AMS then needs to be evaluated. Another option is case-based reasoning, whereby past data are used to understand failure frequency and/or failure pattern recognition (Dinges et al. 2015). This enables an understanding of the machines themselves based on the available data.

When looking into the near future, predictive analytics, remote communications for machines and dashboards, enabling the visualisation of KPIs in AMSs, are seen as the favourable option for service technologies. These technologies support maintenance work, reducing unscheduled repairs, and are enabled through the availability of data from the machines (Dinges et al. 2015).

The value proposition for the customer changes most dramatically when the machine breaks down and needs to be repaired. This is a situation whereby the value capture through ownership of a machine stops abruptly through a breakdown. While ownership of the machine usually brings a general risk, and breakdown is part of the risk of machine ownership, there is still the need to manage the risk and enable the machine to create value for the machine owner again, meaning repair and spare parts' availability. However, this also depends on the business model of the machine owner. If the machine owner is a builder, who works with a digger infrequently, and the machine breaks down on the last day of a building project, the machine may not be used for some time or can easily be exchanged and the repair may not be an important priority. On the contrary, if a harvester breaks down during harvesting season this is likely to be the highest repair priority for the customer. In other words, if the harvester can be repaired, even if the repair is not of a high quality but it enables the machine to work for two more days of the harvesting season, then any larger repairs can be carried out easily, and money, as well as time, does not play an important role. The examples above show that priorities regarding repairs for the customer are different and when the value to the owner, in addition to the cycles of the season and situation, are understood, priorities for repair can be set. When a harvester is broken and needs repair, the worst message that can be given is that the harvester needs to be transported to a workshop, as it cannot be repaired on-site and a larger repair needs to be conducted. Only by understanding the situation (e.g. season, workload or exchangeability) and the business model of the machine owner can an understanding of the way the repair can be conducted, and therefore the highest value for both customer and manufacturer, be gained.

On the contrary, offering good repair business models usually varies across organisations, countries and locations. Hence, the ability to capture real value from repairs cannot often be observed. The problem may be that the repair workshop is often not centrally organised and, hence, the quality of the repair and the extra miles to deliver to the customer are often dependent on the people working in the repair shop. The authors found that communications are considered extremely important by the customer. The quality of the repair and the work to be done are also important aspects; however, communications are ranked highly. This is considered important, as a firm can have many great processes and do better work than its competition, but if it does not communicate what has been done and why, and what the outcome was, then the process is potentially seen as another item on the bill. One example is where a heavy-building machine manufacturer does a general inspection of the status of the machine after its arrival at the repair shop. This includes cleaning the machine and a complete review of it by an expert before work begins. The outcome is an understanding of the status of the machine and whether there are any other aspects of the asset that require attention during the time in the workshop. This process creates value for the customer, as it shows that the manufacturer takes the work seriously. After all, what would be more disturbing if the machine went back to the site after repair and broke down again soon after? Without communication from the repair workshop, however, the customer cannot capture value. A short phone call to the owner of the machine stating that the review has been carried out and that the machine's status is good, or, on the contrary, any information about problems will enable the machine's owner to capture information. The result in the case of negative news will potentially be the offering of additional repairs, which will need to be addressed soon in order for the machine to function fully and reliably. The outcome of such an inspection might also be that there is a major crack in the frame of a digger, or similar, and a repair is wasted money. The decision about an additional repair purchase is then left to the machine owner.

As stated, the speed of delivering repairs as a service can be most important for the customer. At least delivering a decision or a risk assessment will help the customer to make decisions for its business. Information about the time it will take to perform the repair results in decisions such as needing to hire an exchange machine, and so on. One strategic option for a machine manufacturer to enable the customer to capture the maximum value in repairs is to build a service network, which can even be defined as a wider ecosystem. This includes the manufacturers' service organisation, direct certified partners, indirect partners of a lesser certification, competitors and influencers. When the wider ecosystem is well understood and strategically well used, configured and governed, the network can function to the satisfaction of all. When thinking about the value delivery through ecosystem partners, and also how to see competitors as contributors, the value delivery can be maximised for the good of the customer. The complexity of involvement and the number of links to external companies increase. Therefore, the value added needs to be carefully planned and structured.

When combining the ability of the customer and machine manufacturer to conduct good asset management using data from the machines, enabling high-quality maintenance and an understanding of the customer's business models, greater collaboration to create value between the machine manufacturer and owner is possible. Additionally, the machine manufacturer is able to build and sell solutions rather than products. The next section of this chapter will look at the offering of solutions and how the customer can create value for itself by taking on solutions.

3.3 Creating Value from Solutions as a Machine Owner

The first section of this chapter argues that in order to deliver a service, the end value needs to be understood. This is the first step to moving closer to the customer and understanding the customer's business model and business better. This provides the opportunity to move closer to the customer's business and to create value by fully understanding the use and optimisation of the customer's equipment. This enables the machine manufacturer to move into a space where, in combination with the machinery, additional services are sold, such as monitoring the equipment, for example, on fuel efficiency, but also predicting failures. The next step for the manufacturer is to sell uptime of the equipment or to sell the entire value generation and guarantee it. The customer has the option to outsource parts or whole processes to the manufacturer and to increase its productivity in doing so. For example, increasing the capability of predictive analytics in an asset management solution for a manufacturer is much easier, as the scale can be reached and the cost shared with multiple customers. This also means that the management interfaces between the manufacturer and the customer need to be well defined in order to work seamlessly.

Another option for the customer to make the most of the solutions is by no longer owning the assets, but rather paying for the outcome. It is then up to the machine manufacturer or the solution provider to optimise the use of the assets and to ensure that they work to their maximum capacity. One example is that the introduction of autonomous trucks in quarries may be complex for the quarry owner but less complicated for the machine manufacturer. If payment is by outcome, rather than by time and effort of the implementation, the risk stays with the manufacturer and the outcome can be leaner processes and higher value for the customer.

In the case above the trucks are longer paid for by the customer, but they remain in the ownership of the manufacturer. The costs for the machine are therefore not on the books of the customer, but there is a clearer cash flow, which can be linked to the outcome of the production.

In conclusion, the customer of services can easily capture value from service offerings at all levels. The focus should be to create high value throughout the lifetime of the machine and enable the manufacturer to create this value. This will take a common understanding of the situation, as well as an ability and interest in innovating in the space of services.

3.4 Thinking About Outsourcing a Process

When the customer outsources a process to another company there are initial questions to be asked. These include: What is the expected gain from process outsourcing? Which process is not core enough to the company's business model to be outsourced, and which process can be outsourced and gain the most in productivity? The latter indeed asks the question about who can be a partner for the process outsourcing in the first place.

The information presented above emphasises the importance of choosing wisely the process that is outsourced. So what are the aspects for consideration? The outsourced process should enable the company to capture more value through what is offered by the outsourcing company than what the company itself can create internally. This is one area where the customer of the outsourcing company can gain from the proffered expertise. The larger the value increase through outsourcing of the process, the better. However, there are also strategic factors to be considered. On the one hand, outsourcing the process should enable the company to follow and focus on its core business. This means choosing a process for outsourcing that is not too close to the core capability of the company. On the other hand, the core process should be reasonably sized, value can be created and, hence, a process can be taken over by the outsourcing company. It is strategically simple to take an outsourcing company or manufacturer who has a strategic advantage, for example, a strong technology base.

One aforementioned example for process outsourcing is for the machine manufacturer to create value by taking the transport and excavation out of the process of a mine and giving it to a machine manufacturer. The above would facilitate the sale of the fleet to the machine manufacturer, and the machine manufacturer could then implement the operation from the start. One advantage for the machine manufacturer is that they are able to use more advanced technologies and should normally have the knowledge to implement them. Therefore, the value creation is a fixed price for the excavation of material and, hence, stability in pricing, but at the same time the creation of value by enabling a focus on the core business while others look at the technology.

The final section of this chapter has focused on value creation and capture for the customer; from the examples and arguments presented it has become clear that it is particularly important to think about value creation and capture when looking at services and to clarify expectations and strategy in getting the most out of the asset's lifetime.

4 Summary

Industrial firms are servitising, hence shifting their product-oriented strategies to services with the principal aim of expanding their growth and market portfolios. Throughout different service offerings the value creation for manufacturers can be increased; however, services also allow the manufacturer to capture more value. This is based on increased revenues, but also on capturing more of the market share and details on the use of the produced machines. The latter information is important for the machine manufacturer to be able to offer advanced services and solutions. Solutions include the movement of the manufacturer into the production cycle of its customer and participating within that cycle in the form of process outsourcing. One example is taking over the process of the excavation of stone in a quarry and transporting the material to the stone crusher. For the manufacturer to do this, it is important to fully understand the customer's market in order to fully assess the risks.

For the customer it can be advantageous to take on advanced services from the machine manufacturer, which will reduce the risk and costs of optimisation projects. The outcome would be for the customer to focus on its core business, for example, trading stones rather than focusing on the operations of excavation or running a service workshop.

The baseline of working together is acting as network partners instead of suppliers and building relationships instead of transactions. The customer then buys outcomes and solutions rather than products and outputs. This also means the exchange of information about the value creation process, for example, the sharing of information from machines or market expectations. This allows the manufacturer, as well as other partners, to create value.

Overall, both customer and manufacturer are able to create and capture value through service offerings. The focus should be on innovating and reducing value slippage on both sides. Hence, one side creates value and the other captures it, without loss on either side.

References

Dinges, V., et al. (2015). *The future of servitization: Technologies that will make a difference*. Cambridge.

IAM. (2012). *Asset management—An anatomy*. Available at: https://theiam.org/what-is-asset-management/anatomy-asset-management.

Salonen, A. (2011). Service transition strategies of industrial manufacturers. *Industrial Marketing Management, 40*(5), 683–690.

Stremersch, S., Wuyts, S., & Frambach, R. T. (2001). The purchasing of full-service contracts: An exploratory study within the industrial maintenance market. *Industrial Marketing Management, 30*(1), 1–12.

Turunen. (2012). *KIBSIS-key results and findings*. Available at: iris.aalto.fi/en/current/events/taijaturunen__aalto.pdf.

Vandermerwe, S., & Rada, J. (1988). Servitization of business: Adding value by adding services. *European Management Journal, 6*(4), 314–324.

Authors' Biography

Florian Urmetzer is working for University of Cambridge, the Cambridge Service Alliance and is a senior member in the University. His research outcomes are being applied in companies like IBM, Caterpillar and BAE Systems. He has been awarded the European Business Circle (EuBC) outstanding academic research award for research in classical manufacturing companies.

Andy Neely is Pro-Vice-Chancellor for Enterprise and Business Relations at the University of Cambridge, Head of the Institute for Manufacturing (IfM) and Head of the Manufacturing and Management Division of Cambridge University Engineering Department. He is a Fellow of Sidney Sussex College and Founding Director of the Cambridge Service Alliance external link.

Veronica Martinez is a Fellow in the University of Cambridge and a visiting professor at Karlsruhe Institute of Technology, Germany. Priorly she worked at the Cranfield and Strathclyde Universities as researcher and lecturer. Her major research interests revolve around the fields of strategic service performance, value creation and performance measurement and management systems.

Part II

Engineering Network Innovation and Optimisation

5

Engineering Design and Innovation in a Global Context

Dr. Thomas P. Taylor, Dr. Erik Søndergaard, Dr. Tim Minshall, Dr. Letizia Mortara and Prof. Saeema Ahmed-Kristensen

1 Introduction

In the drive for high value engineering, traditional engineering design, product development and innovation processes are evolving. A large number of companies have long since outsourced or offshored their

Dr. T.P. Taylor (✉) · Dr. E. Søndergaard
Technical University of Denmark, Lyngby, Denmark
e-mail: ttay@dtu.dk

Dr. E. Søndergaard
e-mail: esso@dtu.dk

Dr. T. Minshall · Dr. L. Mortara
University of Cambridge, Cambridge, UK
e-mail: thwm100@eng.cam.ac.uk

Dr. L. Mortara
e-mail: lm367@cam.ac.uk

Prof. S. Ahmed-Kristensen
Imperial College London, London, UK
e-mail: s.ahmed-kristensen@imperial.ac.uk

Y. Zhang and M. Gregory (eds.), *Value Creation through Engineering Excellence*,
https://doi.org/10.1007/978-3-319-56336-7_5

production facilities. More recently, trends amongst a number of companies have been observed to outsource and offshore the earlier and more valuable phases in the engineering value chain, including development. This chapter examines the global product development process, which for many companies is a very much learning-by-doing approach, and examines how to support decisions to globalise parts of the development process and monitor the impact of these decisions. Understanding the impact of these decisions can lead to a stronger understanding of how to create high value engineering, whereas as global product development is examined here very much from the perspective of a firm as the key stakeholder whilst outsourcing or offshoring parts of their development, the role of multiple stakeholders is recognised in open innovation processes taking place between stakeholders, in particular within high value engineering networks. These are discussed here.

2 Open Innovation

2.1 Introduction to Open Innovation

The successful creation and capture of new value with global high value engineering networks relies upon the effective collaboration between multiple stakeholders. The complex and dynamic nature of these networks, coupled with the systemic characteristics of many emerging technologies, means that no single company is likely to be able to manage the process of successful innovation on its own. One model that has emerged to describe and support the collaborative approaches required to innovate successfully within global high value engineering networks is *open innovation*.

2.2 How Open Innovation Approaches Can Support Value Creation and Capture for High Value Engineering

As has been widely documented (Chesbrough 2003b; Chesbrough et al. 2006; Huizingh 2011) a number of factors aligned to change the way in which companies innovate over the past two decades. By observing and reflecting on these factors and the lessons provided by the examples of success and failure in leading corporations, a coherent model for describing and operationalising this emergent approach to innovation—*open innovation*—was articulated by Chesbrough (2003b). As part of this model, Chesbrough indicated that firms should on one side access and absorb external knowledge, combine it with internal knowledge to produce innovation (inbound process). At the same time, firms should consider a variety of potential outlets, also beyond the traditional route to market, for the exploitation of the innovation (outbound process).

The management of many leading firms recognised that open innovation presented a potential solution to address the diminishing competitiveness of their current innovation infrastructure and used open innovation as the model around which they could transform their approach to innovation.

The shift towards openness in innovation had been gradual, and examples can be traced back long before the publication of Chesbrough's initial articulation of the open innovation model in 2003 (Di Minin et al. 2010). However, the widespread diffusion of the core open innovation concept gave firms an explicit model for planning, communicating and implementing open approaches to innovation (Mortara and Minshall 2014). Historically, firms who developed open approaches to innovation prior to the publication of the book were led by an 'effectual' decision process (Sarasvathy 2001), i.e. as a result of particular contingencies such as the need to respond to major crises, many forms of openness were experimented with to help the companies survive. The articulation of the open innovation model in 2003 provided some firms with a language to describe activities already ongoing. Other firms were thus able to direct efforts to establishing and deploying open innovation programmes based on Chesbrough's model and visible examples of other firms. Approaches to the implementation of open innovation can therefore be seen as having two phases—one characterised by effectual implementation logic and the other by causal implementation logic—separated by a 'discontinuity' in 2003 (Mortara and Minshall 2014).

2.3 Challenges that Need to be Overcome When Using Open Approaches

There are many potential enablers and obstacles for the successful implementation of open innovation (Mortara and Minshall 2014). Here, in the context of high value engineering, we particularly consider three:

- The implementation of the necessary elements which enable firms to implement the new open innovation routines (e.g. culture change incentives).
- Considerations regarding how distance (geographical or mental) between organisations might impact on the flow of knowledge between organisations.
- The role change of particular agents, such as universities, in the innovation ecosystem.

Culture is most often seen as a barrier in the adoption of open innovation in large companies but has also been identified as an enabling factor (Mortara and Minshall 2011b). Lichtenthaler and Ernst (2006) defined six different attitudes which could distort, act as barriers or overplay the importance of open innovation, of which the 'Not-Invented-Here' (NIH) syndrome (Katz and Allen 1982) is the most frequently cited one. Authors have highlighted a range of issues to overcome the cultural barriers: several authors have highlighted the role of demonstrator projects as important enablers for the acceptance of open innovation (Chiaroni et al. 2011; Westergren and Holmström 2012). Westergen and Holmström (2012) showed that the building of trust with external partners supported the implementation of initial open innovation projects, and thus provided a demonstrator for further collaborative activities. Mortara et al. (2010a) noted the role of 'champions' to support different groups within the firm with different open innovation approaches, providing specific types of motivators in accordance with underlying subcultures. Particular emphasis was placed on the delivery of skills (Mortara et al. 2009) and the creation of boundary spanning objects (Tushman and Scanlan 1981; Fleming and Waguespack 2007). Control mechanisms such as incentives were found to have a positive impact on external (outbound) (Persson 2006) and internal (Minbaeva 2005) knowledge transfer and on the search performance (Salge et al. 2012). Whilst a firm's culture is clearly important, national (Savitskaya et al. 2010) (Chesbrough and Crowther 2006) and regional (Tödtling et al. 2011) cultures may also impact on open innovation implementation.

Within the context of high value engineering networks, the specific location of open innovation-related resources, activities and organisations needs particular consideration. Location in the context of open innovation can be considered in terms of (a) absolute geographic location; (b) proximity to a specific resource; and (c) an organisation's position in a network (Minshall et al. 2014). An organisation's *geographic location* determines the system of innovation within which it operates, and that may qualify it to take part in certain activities (e.g. eligibility to apply for certain regional funding and innovation support programmes) (Edquist 2005) and access other infrastructure elements (e.g. engagement with local universities and colleges) (Huggins 2008; Karlsson 2008). *Proximity* can be viewed in two ways: as relative geographic location or spatial distance, and as relative organisational/cultural compatibility or cognitive distance (Asheim and Gertler 2005; Moodysson et al. 2006; Huggins 2008). Cognitive distance is considered to be more important than spatial distance for knowledge transfer, assimilation and application (Asheim and Gertler 2005). Finally, a firm's *position in a network* can both enable and constrain opportunities for access to external knowledge and new markets and influences the likelihood of knowledge received being novel (Powell and Grodal 2005). Moreover, it determines the relational/social assets and capabilities it can create or gain access to (McEvily and Zaheer 1999). How a high value engineering firm engages with resources and activities to support its open innovation at a particular location can be considered in terms of whether these resources and activities can be accessed remotely, need to be attracted to be closer to the firm, or whether the firm needs to relocate some aspect of its operations closer to the resources or activities (Mortara and Minshall 2014).

One important set of location-related resources for supporting open innovation in the context of high value engineering firms relates to universities. The past 20 years have seen universities become increasingly important components of science and innovation policies in many nations, with growing pressures from government and industry for them to become increasingly strategic actors in processes of innovation and economic development (Deiaco et al. 2012). Indeed, universities have been evolving to become more deeply and strongly linked into the innovation system, and more directly engaged in processes of innovation (Perkmann and Walsh 2007; Youtie and Shapira 2008).

Universities have the potential to contribute to high value engineering firms' innovation processes directly and indirectly (Cohen et al. 2002; Laursen and Salter 2004). Companies can approach universities through different, often institutionalised channels for collaboration. Possible collaboration mechanisms include accessing intellectual property offered by university technology transfer offices, building long-term research collaborations with university departments, sponsoring student activities, or through involvement of spin-off companies based on university research (Chesbrough 2003a; Youtie and Shapira 2008; Brezntiz and Feldman 2012). However, engaging with universities globally as part of a coherent high value engineering network strategy is a capability that not all firms may possess, and thus may require dedicated efforts to develop (Minshall et al. 2015).

2.4 Summary of Processes/Tools That Can be Used to Support Open Innovation with Different Partners Relevant to Value Creation and Capture in a HVE Context

Open innovation provides high value engineering firms with the ability to create and capture new value via networks. However, as shown above, successful use of open approaches to innovation rests upon the firm's ability to overcome certain barriers, and these barriers may be linked to a wide range of factors including:

- The role of culture and, in particular, NIH syndrome (Katz and Allen 1982) in the implementation of open innovation (Mortara and Minshall 2011a).
- Understanding the role of internal R&D capacity (Cassiman and Veugelers 2006; Berchicci 2013) and its links to absorptive capacity (Bogers and Lhuillery 2011).
- Adopting specific management practices to support the implementation of open innovation (Salge et al. 2012).
- Targeting the use of IT systems, as seen in the experience of P&G (Dodgson et al. 2006) and Italcementi (Chiaroni et al. 2011). IT infrastructure is seen as a moderator for open innovation as it helps enable communication across boundaries and networks (Boscherini et al. 2010) and as an element of control (Kuschel et al. 2011).
- Applying specific management tools (Griffiths et al. 1998), taxonomies (Di Minin et al. 2010) or 'watch lists' (Tao and Magnotta 2006; Mortara et al. 2010b) can also be used to find the balance between what to do openly or internally.
- Adopting appropriate location-related activities for different open innovation strategies, i.e. moving operations to specific locations versus attracting organisations to move closer (Minshall and Mortara 2016).
- Understanding the impact of the adoption of 'virtual' platforms for carrying out specific innovation activities across networks (Bughin et al. 2008).
- Appreciating that a change of strategy (such as a shift to a more open approach to innovation across a high value engineering network) is often linked to leadership, the political climate and the internal dynamics of power need to be viewed as moderators in the adoption of open innovation (Pye and Pettigrew 2006).

3 Global Product Development

The establishment of global production sites in low-cost regions is a key force in inducing a more recent trend in Western manufacturing companies—the global distribution of product development (PD) activities. There are many terms in the literature to describe this trend such as the internationalisation of PD, distributed PD or a form of virtual, collaborative PD. In this chapter, we refer to the trend as global product development (GPD), which involves the globalisation of tasks and activities throughout the PD process, from the early concept development stage and detail design through to the final testing of prototypes before production.

The transformation from collocated, cross-functional PD to GPD represents a major transformation in industry today, and companies face the difficult decision of which PD tasks to keep in-house, and which tasks to distribute to independent foreign providers. The following sections outline some of the key impacts associated with GPD, the current practice for decision-making and conclude with recommendations towards a structured decision-making process in GPD.

3.1 The Impacts on the Product Development Process

In engineering design literature there are two terms often used to describe the different sourcing modes in GPD, namely: *offshoring*, the company expands PD to foreign countries whilst maintaining full ownership and control of the subsidiary, and *outsourcing*, the company hands over specific tasks and activities during PD to independent foreign providers. Different sourcing modes have been found to apply at different stages in the PD process. For example, during several case studies Hansen and Ahmed-Kristensen (2011) observed that low value adding activities at the back end of the PD process, such as testing of prototypes before production, were typically outsourced and high value adding activities, such as conceptual development, were typically offshored. The decision to outsource or offshore parts of PD is often driven by the opportunity to reduce development costs. In a survey conducted by the Aberdeen Group (2005), 78% of the 125 manufacturing companies pursued GPD as a strategic decision to reduce PD costs by utilising low-cost, skilled engineers distributed globally. Several studies reaffirm cost reductions as a key motivation for GPD whilst also highlighting less tangible drivers, such as increased access to new competencies and expertise, increased customer base and a reduction in proximity to global markets (Eppinger and Chitkara 2009). Despite the potential benefits associated with GPD the migration from conventional PD, which typically involves the coordination of collocated, cross-functional engineering teams to a form of GPD, which involves the coordination of engineering teams that are globally dispersed and culturally diverse, does not come without challenges. Whilst physical proximity can reinforce social similarity, shared values and expectations, the distance between engineering team members can lead to significant declines in communication and interaction. However, recent studies indicate how companies may have underestimated the challenges with GPD (Eppinger and Chitkara 2009). For example, during their study on collaborative PD in UK manufacturing firms, Littler et al. (1995) highlight how the time required to coordinate GPD increased in comparison with local, cross-functional PD, and the maintenance of the collaborations became the prime objective rather than the development of the product itself. During several case studies, Hansen and Ahmed-Kristensen (2011) found that in an environment where the distance between teams was increased and frequent, spontaneous interactions were reduced; complex development tasks became more difficult to manage and resulted in cultural misunderstandings, design rework and project time delays. Furthermore, the companies were found to switch between outsourcing and offshoring modes, adopting a learning-by-doing approach to GPD, with decisions being made on an ad hoc basis as the collaborations progressed. Given that GPD is a more recent trend in relation to the globalisation of production, it is likely that decision-makers have limited experience prior to embarking on GPD. However, the studies highlight how implementing solutions to challenges on an ad hoc basis can be costly.

The studies exemplify some of the positive and negative impacts of GPD. The learning-by-doing approach indicates the uncertainty companies face when embarking on GPD and highlights the need for a better understanding towards decision-making in GPD. More specifically, the need to support management to make more *informed* decisions during GPD, rather than those that are ad hoc, has been highlighted in the literature (Eppinger and Chitkara 2009; Westphal and Sohal 2012).

3.2 The Role of Information in Strategic Decision-Making

There are four main approaches discussed in the literature for decision-making, namely rationality, bounded rationality, politics and power and the garbage can paradigm (Eisenhardt and Zbaracki 1992). Since there is a need for more informed decision-making and the rational decision approach is founded in information-based decisions, this section focuses on the rational decision-making in the context of outsourcing and offshoring PD. Rational decision-making is characterised by situations whereby actors enter decision situations with known objectives that determine the value of possible consequences of an action (Eisenhardt and Zbaracki 1992). In other words, the approach assumes that there are clearly defined goals for the decision, and that it is possible to systematically choose between different options based on reasons and facts. During Citroen's (2011) investigations of the role of information during strategic decision-making, he concludes that decision-making processes rely heavily on both internal information, such as documentation from the firms' intranet, and external information, such as market information. Decision processes are likely to be influenced by increased availability of information throughout the process. In another study investigating decision-making in design and engineering outsourcing, Shishank and Dekkers (2013) claim that existing decision-making frameworks do not take into account the characteristics of design and engineering, where information can often be incomplete and inaccurate, and the frameworks fail to acknowledge that data only become available progressively as the experiences with outsourcing and offshoring increase. Given that GPD is a relatively recent trend, the opportunity to draw on previous experience is reduced and the availability of information is often limited and hence, the notion that decision-making processes should rely heavily on the availability of information is somewhat paradoxical in the context of GPD.

To summarise, rational decision-making presupposes that information is available (Citroen 2011), whilst GPD decisions are characterised by their high degree of uncertainty, limited available information and often rely on previous experience. Therefore, there is a need to better understand the types of information required by management when making decisions in GPD to better support the decision-making process.

3.3 Decision-Making in Global Product Development

In comparison with areas such as manufacturing and IT services, decision-making in GPD is a relatively unexplored topic. However, during several case studies with Danish manufacturing companies, over 50 decisions were observed when outsourcing and offshoring parts of PD (Søndergaard and Ahmed-Kristensen 2016). The key findings are briefly outlined here:

- The broad variety of information sources used at the companies indicated the complexity of decision-making in GPD, and the information used was often context specific. For offshoring decisions, existing global footprint and market information were key information sources used to support the decisions. For outsourcing decisions, control over activities, core competencies and product/process requirements were key information sources used to support the decisions.
- Most of the studied companies revealed a lack of methods in place for supporting the decision-making process, and hence, the majority of decisions were made on an ad hoc basis. This was particularly common for decisions related to location of new development sites and decisions related to setting up distributed PD teams.

The unstructured and ad hoc manner in which the decisions were made during the study implies that there is a requirement to support the decision-making process in GPD. Decisions made in such an ad hoc manner have been characterised as sporadic decisions by Cray et al. (1988) and often result in an informal and lengthy decision process, which was also reflected in the studied companies.

The findings from the study highlight the need for a more structured approach to strategic decision-making in GPD that supports management and decision-makers in identifying relevant information for specific decision types and encourages the use of practical methods that support implementation of the decisions.

3.4 Recommendations for the Development of a Decision-Making Process in Global Product Development

Based on their study investigating decision-making in GPD, Søndergaard and Ahmed-Kristensen (2016) provide recommendations towards the development of a structured decision-making process. The recommendations aim to support a shift from the observed ad hoc decision process to a more *informed* decision process, and as such, the input of information to support the specific decision types is critical. The information inputs during the decision-making process are derived from two sources: (1) the information that the decision-makers put into the decision and (2) the empirical findings from their study investigating decision-making in GPD (Søndergaard and Ahmed-Kristensen (2016). Adopting this approach enables individualised decisions to be made based on the knowledge and experience of the decision-makers, whilst also incorporating the information and methods used for the specific decision types identified in the previous section. Five recommendations for the development of a structured decision-making process for GPD are outlined:

1. *Decision definition* the decision-maker(s) identifies and forms agreement in relation to the key issues that need to be addressed, i.e. which PD activities to outsource and which to offshore.

2. *The key drivers* the key motivation(s) for the decision is identified, i.e. reductions in PD costs, access to new competencies, or closeness to global production facilities. Examples of motivations from previous studies investigating GPD (Søndergaard and Ahmed-Kristensen 2016) are provided to support this process.
3. *Scenario development* key scenarios describing the potential consequences of the decision are mapped, and these support decision-makers in highlighting any missing information for making the decision. This allows decision-makers to proactively identify potential challenges and hence develop precautionary strategies to avoid deviations.
4. *Uncertainty reduction* methods to support the implementation of the decision are identified. Key methods identified during Søndergaard and Ahmed-Kristensens study (2016), such as risk assessments, design reviews or vendor selection methods, are provided to support this process. However, decision-makers can also apply other methods that are deemed appropriate for the specific context and decision.
5. *Decision action* a strategic action plan is developed, including the key drivers for the decision, any missing information and potential challenges as a result of the decision and the methods adopted to support the implementation of the decision.

Once the decision is made, it is important to monitor and measure the impacts that outsourcing or offshoring parts of PD have at the company to encourage an organisational learning approach to GPD. Monitoring the impacts enables management to make adjustments along the process and hence, avoid any deviations as a result of common challenges encountered during GPD. The following sections focus on the development of performance measures in the context of GPD.

4 Performance Measurement

Although performance measurement (PM) is a well-established concept for business processes in general, with several comprehensive studies exemplifying PM as a practical tool to support decision-making (Kaplan and Norton 1996; Neely et al. 2000), there has been less focus towards PM in the context of GPD. The following sections aim to address this by adopting theoretical concepts from PM system design to investigate PM in the context of GPD. A framework to support the development of key performance indicators in GPD is developed, and its application is exemplified in two Danish manufacturing companies.

4.1 Performance Measurement System Design

Performance is defined as the effectiveness and efficiency of a process with the purpose of achieving a fixed objective or set of goals (Neely et al. 2000). The measurement of performance requires a PM system, with the critical element being a balanced set of financial and non-financial key performance indicators (KPIs), which in this chapter are defined as quantifiable metrics that help an organisation measure the success of critical factors. Given the centrality of KPIs in a PM system, ensuring purposeful and measurable KPIs are developed is critical. During their work on PM system design, Neely et al. (2000) propose six desirable characteristics for developing successful KPIs:

1. Indicators should be derived from the company's strategy.
2. The purpose of the indicator must be made explicit.
3. Data collection and methods of calculating performance must be clear.
4. All stakeholders must be involved in the selection of the indicators.
5. The indicator should take account of the organisation.
6. The indicators should change as circumstances change.

Although the characteristics are intended for the development of KPIs for business processes in general, they also offer indication towards what constitutes a 'good' KPI for PM in environments such as GPD.

4.2 Key Performance Indicators: Leading and Lagging

During the development of the balanced scorecard, Kaplan and Norton (1996) identified two types of KPIs important for performance measurement:

- Lagging KPIs: that measure output of past activity and typically consist of financial indicators.
- Leading KPIs: that measure factors influencing a process and are drivers of performance.

Lagging KPIs (outcome measures) without leading KPIs (performance drivers) do not communicate how the outcomes of a process are to be achieved. For example, Taylor and Ahmed-Kristensen (2016) describe observations at a large Danish pharmaceutical company that were in the process of developing a new syringe. It was communicated at the beginning of the development project that the time to market for the new syringe was critical to avoid the risk of their component suppliers developing and releasing a similar product. Based on this understanding, the project manager and team identified the time taken for documentation approval by the internal approval committee for the project as a key factor that could lead to project time delays and hence, delay the time to market for the product. Therefore, 'The time-taken for document approval' was developed as a leading KPI to monitor this and prompt the team to take action when the expected approval time for project documents was exceeded. Furthermore, 'The number of days delayed due to document approval' was set up as a lagging KPI to measure the impact in relation to the projects time to market.

When developing leading KPIs, Rhodes et al. (2009) state that 'contrary to simple status oriented measures typically used on most projects, leading indicators are intended to provide insight into the probable future state, allowing projects to improve the management and performance of complex programs before problems arise'.

Despite the importance of developing KPIs to evaluate the outcome of a process (lagging KPIs), and KPIs to enable the course of action to be altered (leading KPIs), a recent study investigating PM in GPD highlighted how leading KPIs in particular appeared to be illusive in practice (Taylor and Ahmed-Kristensen 2016). In fact, a general criticism towards KPIs in engineering design is they tend to be lagging in nature and focus on the more tangible outcomes of the PD process such as return on investment or break-even time (Tatikonda 2007). However, measuring the outcome of GPD alone does not provide the predictive insight required to avoid deviations along the process. For example, Taylor and Ahmed-Kristensen (2016) observed that lagging KPIs did not provide feedback regarding challenges encountered during GPD, such as cultural differences and team proximity, which influenced the success and resulted in project time delays. In environments of high uncertainty such as GPD, identifying key challenges and developing leading KPIs accordingly is an important step to support the development of precautionary strategies and hence, avoid deviations along the process.

The following sections outline the development and application of a method to support the development of KPIs for GPD.

4.3 The Development of Key Performance Indicators for Global Product Development

In this section, the KPI Development Toolkit is presented, which was developed by adopting well-established methodologies for PM (Kaplan and Norton 1996; Neely et al. 2000) and utilising key findings from a research project that investigated PM in GPD (Taylor and Ahmed-Kristensen 2016). The KPI Development Toolkit aims to support project managers in manufacturing companies to develop and implement leading and lagging KPIs in GPD. The methodological framework for the toolkit consists of three phases that are carried out for 5 hours facilitated KPI development workshop. The three phases are summarised here.

Phase 1: Key concepts

Aim Develop an understanding towards key theoretical concepts of performance measurement.

Process The facilitator uses examples of best practice to develop an understanding towards the development and application of KPIs; PM to support decision-making; and the relationship between leading and lagging KPIs. It is important that key stakeholders involved with GPD at the company are involved at this phase to ensure the purpose of PM is understood.

Outcome Commitment from the GPD team towards developing and using KPIs.

Phase 2: KPI development

Aim Provide a structured approach for developing both leading and lagging KPIs.

Process The approach for Phase 2 follows the methodological framework illustrated in Fig. 1. There are three levels of PM illustrated in the framework, namely Business-level; Project-level; and Task-level, and coherence between KPIs developed at each level is important to encourage behavioural alignment at the company. The toolkit focuses on developing KPIs at the Project-level where less has been reported in the literature.

The methodological framework contains three key steps. First, key motivations and challenges for the GPD project are identified and prioritised and the cause–effect relationships between (1) key motivations that represent the desired outcome for the project and (2) key challenges that influence the success towards this outcome are identified and mapped to a cause–effect fishbone diagram (Kitcher et al. 2013). Strategies to prevent the influence on success are developed to support the development of quantifiable KPIs. Based on the strategies, leading KPIs are developed to avoid the identified influence on success and support the identification of

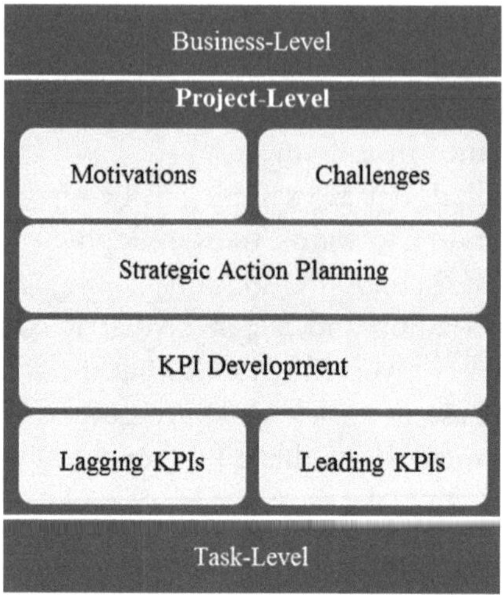

Fig. 1 Methodological approach for the development of leading and lagging KPIs

deviations along the process. Lagging KPIs are developed to measure the achievement towards the desired outcome. The strategies and KPIs developed are mapped to the companies PD process, indicating where along the process the strategies require implementing and the frequency of measurement for the developed KPIs. Phase 2 should involve key stakeholders from the GPD project and be repeated at important intervals during the project to ensure the KPIs change as the circumstances change.

Outcome Critical factors are identified and preventative strategies are developed and aligned with the PD process to support the development of measurable KPIs.

Phase 3: Reporting

Aim Support the documentation and implementation of leading and lagging KPIs.

Process Each of the developed KPIs is documented according to a KPI Template. The criteria in the KPI Template are based on previous templates for documenting KPIs (Neely et al. 2000) to ensure: the purpose and formula for measuring the KPI are understood; the main respondent for the KPI is outlined; and the frequency of measurement and targets are clearly defined. Following this, the KPIs are documented according to a KPI Visualisation Board, which allows for simple monitoring of both leading and lagging KPIs and provides indication towards key challenges, proposed solutions, key achievements and the next steps for measurement. The main respondent for monitoring the KPIs in the GPD project should be involved during Phase 3.

Outcome The KPI Template and KPI Visualisation Board are completed for monitoring the KPIs.

4.4 Application of the KPI Development Toolkit

The key results for the implementation of the KPI Development Toolkit at two large Danish manufacturing companies are presented in the following section (see Table 1 for company characteristics).

A 5-hour KPI development workshop was held at each of the companies with the aim to establish KPIs for the GPD projects. Prior to the

Table 1 Key characteristics of companies for the application of the KPI Development Toolkit

Characteristics	Company A	Company B
Employees	420 (global)	900 (global)
Industry	Air conditioning and refrigeration systems	Satellite communications for aviators
Offshore R&D facilities	China	South Africa
GPD project	Develop a software program to ensure future PD projects follow a standard process	Develop a new radio system for airplanes to improve communication when airborne
No. of KPI workshop participants	6	8

KPI development workshop, the project team from both companies had no experience with developing KPIs, and hence, clarifying key theoretical concepts in relation to leading and lagging KPIs during Phase 1 was particularly insightful. Phase 2 was the most time-consuming phase of the workshop, taking approximately 3 hours to complete. This was primarily due to the difficulties experienced during the prioritisation of key challenge factors in the GPD projects as many of the project team came from different functional backgrounds. For example, in company B the project manager felt the lack of commitment towards standard company procedures was a key issue likely to impact the project delivery time. However, the software engineer felt the lack of alignment between software and hardware development would result in quality issues. Despite the lengthy discussions, Phase 2 of the workshop was undoubtedly the most valuable for both companies as it supported the alignment of expectations within the project teams. Figure 2 illustrates an excerpt from the fishbone diagram completed at company B during Phase 2. Firstly, the project manager derived the desired outcomes for the project based on the organisation-level KPIs at the company. Secondly, the project team agreed that the lack of understanding towards key company procedures was the most critical factor influencing success of

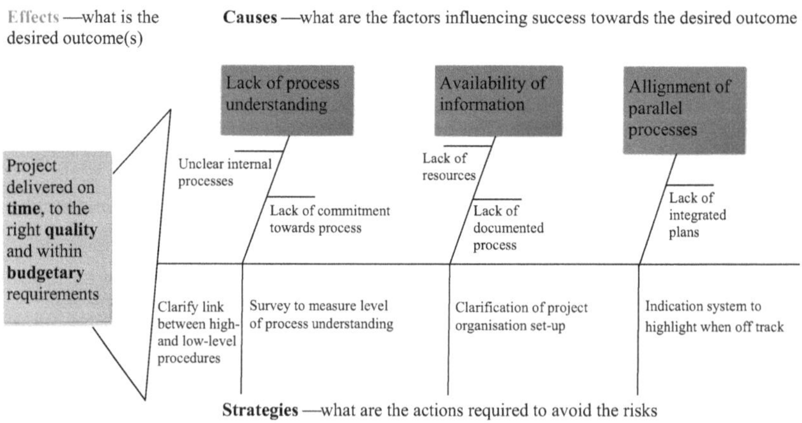

Fig. 2 Example of identifying key cause-effect relations and developing strategies at company B

the GPD project. Finally, it was decided that an internal company survey, focused on testing the level of understanding towards company procedures within the project team, should be developed to identify areas for improvement.

The project team considered the survey a powerful evaluation tool to capture data over time and monitor the level of understanding towards company procedures. However, it was critical the questions in the survey were carefully devised to ensure the results provided a true reflection of understanding towards key company procedures for the project in hand. As such, it was not possible to develop the internal company survey during the KPI development workshop as key decisions regarding the content of the survey needed approval from the quality assurance team. This hindered the opportunity to fully complete the KPI Template in Phase 3 (see Fig. 3). Without the questions in the survey, it was difficult to identify the key targets for the KPI. This was also the case for company A, where the strategy they developed required additional time outside of

Key Performance Indicator:	Level of understanding towards company procedures
Purpose	Increase level of understanding of processes and planning
Action plan	Develop 5 minute survey to tests level of understanding towards company procedures - distribute on regular basis
Critical influence factor related to	Process awareness and adherence at the company
Desired outcome related to	Improve level of understanding and adherence towards common processes supports achieve desired customer requirements, cost and time
Calculation (%of, #of, $of, ...)	Scaling system based on survey - expected % of correct answers Vs actual % of correct answers
Forcasted target (weekly/ monthly)	To be decided
Maximum target (weekly/ monthly)	To be decided
Measurement frequency	Monthly - the results are to be shared within the teams
Product development stage	Continuous—beginning of each stage of the PD plan
Data source	Participant results from survey
KPI type (Leading/ Lagging)	Leading - identifies where there is a lack of understanding towards processes and hence, allows for corrective action along the process
Main responsible	Program management office
Additional comments:	Different members have different understanding towards processes - requirement for answers in survey to be weighted Results from the survey should be populated digitally to ensure a database can be created and analysed over time

Fig. 3 Example of completed KPI template at company B

the 5-hour KPI development workshop, and hence, key targets for the developed KPIs could not be set.

To summarise, although the identification and prioritisation of key cause–effect relationships in Phase 2 was the most time-consuming step in the KPI Development Toolkit, it supported in aligning expectations for the GPD projects across key project stakeholders. Furthermore, designing strategic action plans was an important step to support the development of quantifiable, leading KPIs to measure intangible influence factors. However, the design and implementation of strategic action plans exceeded the time allocated for the KPI development workshop in both company A and company B. Therefore, Phase 3 in the KPI Development Toolkit should not be completed until the strategic action plans are implemented to ensure key targets for the KPIs are determined.

5 Conclusions

The chapter aims to unravel some of the supporting processes to create high value engineering networks through open innovation and through globalising (offshoring and outsourcing) parts of the engineering functions in a product development process. The motivations for the global product developments processes are now relatively well understood in research, yet these bring their own challenges. To support firms in evaluating, selecting and monitoring global product development processes, cases studies were discussed to highlight the sporadic nature of decision-making and the lack of understanding of key performance indicators, which can successfully monitor and address these challenges. Offshoring decisions, which typically are the approach for the earlier and higher value elements of the engineering value chain, were found to be made in an ad hoc manner with complex level of information needed, and a lack of relevant method to support these types of decision added. A framework to monitor the impact of these decisions is presented, focusing on both reducing the risk and monitoring for success, with key performance indicators developed specifically for globalised product development. The importance of creating, for example, the appropriate cultures, appropriate IT infrastructure and understanding the impact of virtual platforms to support open innovation was highlighted.

This chapter contributes to understanding the challenges of open innovation in high value engineering networks; making decisions to globalise the higher value phases of the engineering value chain (product development process); and monitoring the impact of these decisions (global product development processes). For each of these initial recommendations and frameworks are suggested to increase the likelihood to obtain success.

References

Aberdeen Group Inc. (2005). *The global product design benchmark report.* Retrieved from http://www.plm.automation.siemens.com/en_us/Images/Q4_RA_GlobalProductDesign_JB_123005bspf_tcm1023-23397.pdf.

Asheim, B., & Gertler, M. S. (2005). The geography of innovation: Regional innovation systems. In J. Fagerberg, D. C. Mowery, & E. G. Nelson (Eds.), *The oxford handbook of innovation.* Oxford: Oxford University Press.

Berchicci, L. (2013). Towards an open R&D system: Internal R&D investment, external knowledge acquisition and innovative performance. *Research Policy, 42*(1), 117–127.

Bogers, M. S., & Lhuillery, S. (2011). A functional perspective on learning and innovation: Investigating the organization of absorptive capacity. *Industry and Innovation, 18*(6), 581–610.

Boscherini, L., Chiaroni, D., Chiesa, V., & Frattini, F. (2010). How to use pilot projects to implement open innovation. *International Journal of Innovation Management, 14*(6), 1065–1097.

Breznitz, S., & Feldman, M. P. (2012). The Engaged University. *Journal of Technology Transfer, 37,* 139–157.

Bughin, J., Chui, M., & Johnson, B. (2008, June 08). Next steps in open innovation. *The McKinsey Quarterly*, 1–8.

Cassiman, B., & Veugelers, R. (2006). In search of complementarity in innovation strategy: Internal R&D and external knowledge acquisition. *Management Science, 52*(1), 68–82.

Chesbrough, H. (2003a). The era of open innovation. *Sloan Management Review Spring, 2003,* 35–41.

Chesbrough, H. (2003b). *Open innovation: The new imperative for creating and profiting from technology*. Boston: Harvard Business School Press.
Chesbrough, H., & Crowther, A. K. (2006). Beyond high tech: Early adopters of open innovation in other industries. *R & D Management, 36*(3), 229–236.
Chesbrough, H., Vanhaverbeke, W., & West, J. (2006). *Open innovation: Researching a new paradigm*. Oxford: Oxford University Press.
Chiaroni, D., Chiesa, V., & Frattini, F. (2011). The open innovation journey: How firms dynamically implement the emerging innovation management paradigm. *Technovation, 31,* 34–43.
Citroen, C. L. (2011). The role of information in strategic decision making. *International Journal of Information Management, 31*(6), 493–501.
Cohen, W. M., Nelson, R. R., & Walsh, J. (2002). Links and impacts: The influence of public research on industrial R&D. *Management Science, 48*(1), 1–23.
Cray, D., Mallory, G. R., Butler, R. J., Hickson, D. J., & Wilson, D. C. (1988). Sporadic, fluid and constricted processes: Three types of strategic decision making in organisations. *Journal of Management Studies, 25*(1), 13–39.
Deiaco, E., Hughes, A., & McKelvey, M. (2012). Universities as strategic actors in the knowledge economy. *Cambridge Journal of Economics, 36,* 525–541.
Di Minin, A., Frattini, F., & Piccaluga, A. (2010). FIAT: Open innovation in a downturn (1993–2003). *California Management Review, 52*(3), 132–159.
Dodgson, M., Gann, D., & Salter, A. (2006). The role of technology in the shift towards open innovation: The case of Procter & Gamble. *R&D Management, 36*(3), 333–346.
Edquist, C. (2005). Systems of innovation: Perspectives and challenges. In J. Fagerberg, D. C. Mowery, & R. Nelson (Eds.), *The Oxford Handbook of Innovation*. Oxford: Oxford University Press.
Eisenhardt, K., & Zbaracki, M. (1992). Strategic decision making. *Strategic Management Journal, 13,* 17–37.
Eppinger, S. D., & Chitkara, A. R. (2009). The practice of global product development. *MIT Sloan Management Review, 47*(4), 22–30.
Fleming, L., & Waguespack, D. M. (2007). Brokerage, boundary spanning, and leadership in open innovation communities. *Organization Science, 18*(2), 165–180.
Griffiths, D., Boisot, M., & Mole, V. (1998). Strategies for managing knowledge assets: A tale of two companies. *Technovation, 18*(8–9), 529–539.

Hansen, Z. N. L., & Ahmed-Kristensen, S. (2011). Global product development: The impact on the product development process and how companies deal with it. *International Journal of Product Development, 15*(4), 205–226.
Huggins, R. (2008). The evolution of knowledge clusters. *Economic Development Quarterly, 22*(4), 277–289.
Huizingh, E. K. (2011). Open innovation: State of the art and future perspectives. *Technovation, 31*(1), 2–9.
Kaplan, R. S., & Norton, D. P. (1996). Linking the balanced scorecard to strategy. *California Management Review, 39*(1), 53–79.
Karlsson, C. (2008). *Handbook of research on innovation and clusters* (pp. 1–16). Cheltenham: Edward Elgar Publishing.
Katz, R., & Allen, T. J. (1982). Investigating the not invented here (NIH) syndrome: A look at the performance, tenure, and communication patterns of 50 R&D project groups. *R & D Management, 12*(1), 7–19.
Kitcher, B., McCarthy, I. P., Turner, S., & Ridgway, K. (2013). Understanding the effects of outsourcing: Unpacking the total factor productivity variable. *Production Planning & Control: The Management of Operations, 24*(5), 308–317.
Kuschel, J., Remneland, B., & Kuschel, M. H. (2011). Open innovation and control: A case from Volvo. *International Journal of Networking and Virtual Organisations, 9*(2), 123–139.
Laursen, K., & Salter, A. (2004). Searching high and low: What type of firms use universities as a source of innovation? *Research Policy, 33*(8), 1201–1215.
Lichtenthaler, U., & Ernst, H. (2006). Attitudes to externally organising knowledge management tasks: A review, reconsideration and extension of the NIH syndrome. *R & D Management, 36*(4), 367–386.
Littler, D., Leverick, F., & Bruce, M. (1995). Factors affecting the process of collaborative product development: A study of UK manufacturers of information and communications technology products. *Journal of Product Innovation Management, 12*(1), 16–32.
McEvily, B., & Zaheer, A. (1999). Bridging ties: A source of firm heterogeneity in competitive capabilities. *Strategic Management Journal, 20*(12), 1133–1156.
Minbaeva, D. B. (2005). HRM practices and MNC knowledge transfer. *Personnel Review, 34*(1), 125–144.
Minshall, T., & Mortara, L. (2016). Evolving physical infrastructure to support open innovation in the digital age: Case studies from the UK. In *3rd Annual World Open Innovation Conference.* Barcelona.

Minshall, T., Kouris, S., Schmithausen, P., Mortara, L., & Weiss, D. (2014). Developing infrastructure to support open innovation: Analysis of case studies in the East of England. *International Journal of Innovation & Technology Management, 11*(1), p. 1440006. See http://www.worldscientific.com/doi/abs/10.1142/S0219877014400069.

Minshall, T., Mortara, L., & Ulrichsen, T. (2015). University–industry partnerships and open innovation implementation: Experiences from Cambridge. In *2nd World Open Innovation Conference 19–20th November 2015.* Santa Clara, CA.

Moodysson, J., Coenen, L., & Asheim, B. (2006). Explaining spatial patterns of innovation: Analytical and synthetic modes of knowledge creation in the Medicon Valley life-science cluster. *Environment and Planning A, 40*(40), 1040–1056.

Mortara, L., & Minshall, M. (2011a). How do large multinational companies implement open innovation? *Technovation, 31*(10–11), 586–597.

Mortara, L., & Minshall, T. (2011b). How do multinational corporations implement open innovation? *Technovation, 31*(10–11), 586–597.

Mortara, L., & Minshall, T. (2014). Patterns of implementation of open innovation in multinational corporations. In H. Chesbrough, W. Vanhaverbeke, & J. West (Eds.), *New Frontiers in Open Innovation.* Oxford: Oxford University Press.

Mortara, L., Mills, J., Shawcross, J., & Minshall, T. (2009). Skills for open innovation. *R&D Management Association (RADMA).*

Mortara, L., Slacik, I., Napp, J. J., & Minshall, T. (2010a). Implementing open innovation (OI): Cultural issues. *International Journal of Entrepreneurship and Innovation Management, 11*(4), 369–397.

Mortara, L., Thomson, R., Moore, C., Armara, K., Kerr, C., Phaal, R., et al. (2010b). Developing a technology intelligence strategy at Kodak European Research: Scan and target. *Research—Technology Management, 53*(4), 27–38.

Neely, A., Mills, J., Platts, K., Richards, H., Gregory, M., Bourne, M., et al. (2000). Performance measurement system design: Developing and testing a process-based approach. *International Journal of Operations & Production Management, 20*(10), 1119–1145.

Perkmann, M., & Walsh, K. (2007). University–industry relationships and open innovation: Towards a research agenda. *International Journal of Management Reviews, 9*(4), 259–280.

Persson, M. (2006). The impact of operational structure, lateral integrative mechanisms and control mechanisms on intra-MNE knowledge transfer. *International Business Review, 15*(5), 547–569.
Powell, W. W., & Grodal, S. (2005). Networks of innovators. In J. Fagerberg, D. C. Mowery, & E. G. Nelson (Eds.), *The Oxford Handbook of Innovation*. Oxford: Oxford University Press.
Pye, A., & Pettigrew, A. (2006). Strategizing and organizing: Change as a political learning process, enabled by leadership. *Long Range Planning, 39*(6), 583–590.
Rhodes, D. H., Valerdi, R., & Roedler, G. J. (2009). Systems engineering leading indicators for assessing program and technical effectiveness. *Systems Engineering, 12*(1), 21–35.
Salge, T. O., Bohné, T. M., Farchi, T., & Piening, E. P. (2012). Harnessing the value of open innovation: The moderating role of innovation management. *International Journal of Innovation Management, 16*(3), see https://doi.org/10.1142/S1363919612400051
Sarasvathy, S. D. (2001). Causation and effectuation: Toward a theoretical shift from economic inevitability to entrepreneurial contingency. *Academy of Management Review, 26*(2), 243–263.
Savitskaya, I., Salmi, P., & Torkkeli, M. (2010). Barriers to open innovation: Case China. *Journal of Technology Management and Innovation, 5*(4), 10–21.
Shishank, S., & Dekkers, R. (2013). Outsourcing: Decision-making methods and criteria during design and engineering. *Production Planning & Control, 24*(4–5), 318–336.
Søndergaard, E., & Ahmed-Kristensen, S. (2016). *Decision making for global product development*. Ph.D. thesis, Technical University of Denmark, Denmark.
Tao, J., & Magnotta, V. (2006). How air products and chemicals "identifies and accelerates". *Research Technology Management, 49*(5), 12–18.
Tatikonda, M. V. (2007). Product development performance measurement. In C. Loch & S. Kavadias (Eds.), *Handbook of new product development management* (pp. 199–215). Abingdon: Routledge.
Taylor, T. P., & Ahmed-Kristensen, S. (2016). *Performance measurement in global product development*. Ph.D. thesis, Technical University of Denmark, Denmark.
Tödtling, F., Van Reine, P. P., & Dörhöfer, S. (2011). Open innovation and regional culture-findings from different industrial and regional settings. *European Planning Studies, 19*(11), 1885–1907.

Tushman, M. L., & Scanlan, T. J. (1981). Boundary spanning individuals: Their role in information transfer and their antecedents. *Academy of Management Journal, 24*(2), 289–305.
Westergren, U. H., & Holmström, J. (2012). Exploring preconditions for open innovation: Value networks in industrial firms. *Information and Organization, 22,* 209–226.
Westphal, P., & Sohal, A. S. (2012). Taxonomy of outsourcing decision models. *Production Planning & Control, 24*(4–5), 1–12.
Youtie, J., & Shapira, P. (2008). Building an innovation hub: A case study of the transformation of university roles in regional technological and economic development. *Research Policy, 37,* 1188–1204.

Authors' Biography

Dr. Thomas Taylor is a postdoctoral researcher at the Management Engineering department of the Technical University of Denmark. His research is focused on global product development and the design and implementation of performance measurement and decision-making tools to support this process. His research adopts empirical research methods in close collaboration with industry.

Dr. Erik Søndergaard holds a M.Sc and a Ph.D in management engineering from the Technical University of Denmark. His research considers the globalisation of product development and innovation activities. He has worked in close collaboration with industrial partners in developing decision support for the outsourcing and offshoring of product development.

Dr. Tim Minshall is Reader in Technology and Innovation Management at the University of Cambridge, and Head of the Centre for Technology Management and deputy Head of the Institute for Manufacturing in the Department of Engineering. Academic interests include open innovation, university–industry collaborations, technology business investment and incubation, engineering education and digital fabrication.

Dr. Letizia Mortara is Senior Research Associate at the University of Cambridge and a By-Fellow at Churchill College, Cambridge. She is an Associate Editor for the R&D Management journal, and currently, she is a Visiting Scholar at the University of Bologna in Italy and Visiting Professor at Sogang University (South Korea). Her research interests include Technology intelligence (i.e. the activity set-up in order to keep abreast with the latest developments in technology) and Open Innovation. She is currently studying the advent of Digital Fabrication technologies in manufacturing (e.g. 3D printing) and their implications for business.

Prof. Saeema Ahmed-Kristensen is Professor of Design Engineering and Design Methodology of the Dyson School of Design Engineering, Imperial College London. Her research focuses upon improvements of both products (including product service systems) and processes (creative, product development and innovation) through developing a scientific understanding of the processes. She has over 100 reviewed publications and works in close collaboration with industry, from aerospace, oil drilling equipment, medical devices, to consumer products such as headsets.

6

Engineering in International Business Networks: The Motivations and Practices of Chinese MNCs

Quan Zhou, Xiaobo Wu and Yongjiang Shi

1 Introduction

China, as the world's second largest economy, has the most dynamic and fast-growing market for both foreign and domestic companies. It is now the hot spot for every multinational corporation (MNC). In face of the fierce competition from their Chinese counterparts, one way for Western companies to respond is to develop products tailored to the local demand. This can be done in two ways: Localisation of products designed for Western markets, or new products developed specifically for Chinese market. Both

Q. Zhou (✉) · X. Wu
School of Management, Zhejiang University, Hangzhou, China
e-mail: quanzhou@zju.edu.cn

X. Wu
e-mail: xbwu@zju.edu.cn

Y. Shi
Institute for Manufacturing, Engineering Department, Cambridge University, Cambridge, UK
e-mail: ys@eng.cam.ac.uk

Y. Zhang and M. Gregory (eds.), *Value Creation through Engineering Excellence*,
https://doi.org/10.1007/978-3-319-56336-7_6

require a strong presence of engineering capabilities (especially research, design and development capabilities, i.e. R&D) locally. This might explain the growing number of engineering facilities (e.g. R&D centres) invested by MNCs in China. However, previous research on international R&D usually focuses on large MNCs from developed countries internationalised R&D to another developed country, usually among the USA, Europe and Japan (Florida 1997; Asakawa and Som 2008). Only in the last decade, with the structural change in China's policy (e.g. entry to WTO) and economy, Western companies start to explore China as a destination for R&D investment (von Zedtwitz 2004). Within these 10 years, numerous MNCs have set up their R&D units in China. According to UNCTAD (2005), China as a destination for R&D is more attractive than the USA. Specifically, a recent report estimates that there are about 1600 foreign R&D units in China (Financial Times 2012), usually concentrated on the big cities such as Beijing, Shanghai, Guangzhou and Shenzhen (Sun and Wen 2007; von Zedtiwitz 2004).

In the meantime, it is also worth noting that Chinese companies are encouraged by the government's "going out" strategy, which suggests Chinese companies to play a part in international capital market and invest overseas (Hong and Sun 2006). Chinese companies such as Huawei, Lenovo and Haier are now operating on a global scale in all the value chain activities (R&D, production and sales). Earlier research however, mainly focused on Chinese companies' FDI from a macroeconomic level. Very little research has been done in investigating in detail the characteristics of China's R&D activities overseas. This may be because R&D is a universal function and therefore less location specific (Motohashi 2012). We argue that while numerous papers have addressed R&D internationalisation, they are mostly based on evidences and experiences from Western MNCs. If we can recognise the differences between Chinese and Western MNCs, one might argue that Chinese companies will behave differently from their Western counterparts.

To address the gaps in both academic research and practical needs, this paper reports our findings from investigating the following questions:

- Why do Chinese MNCs conduct R&D overseas?
- How is the R&D location selected and premise established?

- What are the unique challenges for Chinese MNCs managing R&D out of China?

Western MNCs have been conducting R&D in China for decades. The challenges and barriers they have encountered in China might be similar to Chinese companies extending R&D to the West, as these challenges and barriers are caused by the same contradiction: The differences in economic system, economic growth speed, size of the market and technological innovation capabilities. We believe it is worthwhile to compare and contrast the differences of international R&D activities conducted by Western and Chinese companies, and it may lead to fresh insights. In response to the research questions, we selected six cases to study R&D units that were set up by companies from Europe and China.

2 Literature Review

Two waves of international R&D activities have been captured in the literature. In the first wave, MNCs from developed country invest R&D aboard looking for new technologies or complementary assets that can maintain or further their global competitiveness (Buckley and Casson 1998; Florida 1997). These R&D investment activities normally happen between two industrialised countries. With the uprising of developing countries such as China and India, the second wave of international R&D concerns about the R&D investment from developed countries to developing countries. The strategic intention is to leverage the technological advancement in MNCs' home countries to the host countries where engineering capabilities are relatively weak (Asakawa and Som 2008; von Zedtwitz 2004). In recent years, especially after the financial crisis, international investments from Chinese companies are starting to grow, among which R&D investment is also included. This can hardly be labelled as a third wave as it is not a global phenomenon yet. But for Chinese firms, overseas R&D investment is a very important route for acquiring technology that could help them compete in both domestic and foreign markets. We will review the R&D internationalisation literature from four perspectives: motivations, location choice, organisational structure and managerial issues.

2.1 Motivations and Location Choice

In an early attempt to address the objectives of international R&D activities, Bartlett and Ghoshal (1990) provide a topology based on the original location of innovation and the location of product market. Two typical models are: "centre-for-global", that is to create a new product or process at home country for the global market, and "local-for-local", which refers to local subsidiaries creating their own innovation in response to the needs of the local demands. Two other models are "locally leveraged", in which innovations are developed locally for global use, and "globally linked", in which globally networked R&D units collaboratively develop products or process for the world market.

Kuemmerle (1999) suggests a simpler but widely cited classification. He distinguishes "home-based-augmenting" R&D units with the objectives of transferring locally created knowledge to a central R&D unit from "home-based-exploiting" R&D units that transfer knowledge created in the central unit to the overseas units.

Le Bas and Sierra (2002) propose four locational strategies based on a firm's existing capabilities and home/host countries technological profile. The first strategy is "technology seeking": A company can invest R&D in a host country with proven strong technology capabilities to offset the home country weakness. The second strategy is exact the opposite: The asymmetry of technological capabilities is reversed and MNCs use "home-based-exploiting" strategy to exploit their advanced technology in a region weak in the field. The third strategy is "home-based-augmenting", in which both home country and host country are strong in a technological field, and R&D activities in the host country are set to follow technology development or acquire complementary capabilities. The fourth strategy is "market seeking". The main driver is not the advancement of technology in home or host country but an international expanding option.

The decision of selecting a particular location for overseas R&D units usually involves higher management, R&D department and strategy department (von Zedtwitz and Gassmann 2002). These motivational factors can be classified into six categories (Gammeltoft 2006): market driven, production driven, technology driven, innovation driven, cost driven and policy driven. Table 1 provides a detailed description of Gammeltoft's classification, and Table 2 adopts Gammeltoft's classification to summarise motives for R&D internationalisation in the literature.

Policy-driven motives are also important in the context of R&D in China (von Zedtwitz 2004). Locating R&D in China, particularly in Beijing, is important to keep continuous communication with the Chinese government and standard shaping bodies. For example, one important motivation for telecommunication companies to operate a Beijing R&D centre is to be involved in mobile telecommunications standard setting activities by the government (Motohashi 2012).

Reducing cost is a strong motive for MNCs performing international manufacturing. It is also a reason for foreign firms to do R&D in China (Gassmann and Han 2004; Motohashi 2012; von Zedtiwitz 2004). This is later discussed in the case studies that it is not cheap to do R&D in China, especially with limited number of experienced research staffs.

Table 1 Motives for R&D internationalisation. Adapted from Gammeltoft (2006)

Motives	Activities involved
Market driven	Exploit existing company-specific assets more widely; motivated by market size and proximity; support local sales, closeness to lead customer, improve responsiveness in terms of both speed and relevance
Production driven	Supporting local manufacturing operations
Technology driven (pull)	Tapping into foreign S&T resources, technology monitoring (especially competitor analysis), acquire/monitor local expertise, knowledge and technologies
Innovation driven (push)	Generating new company-specific assets; attaining a faster and more varied flow of new ideas, products and processes; capitalise on location-specific advantages through an international division of labour between R&D laboratories
Cost driven	Exploiting factor cost differentials
Policy driven	National regulatory requirements or incentives, tax differentials, monitoring and exploitation of regulations and technical standards

Table 2 A comparison of motives found in the literature

Motives	General international R&D	Foreign R&D in China	Chinese R&D in West countries
Market driven	Cantwell and Mudambi (2005), Gammeltoft (2006), Gassmann and von Zedtwitz (2002), Gerybadze and Reger (1999), Kuemmerles (1999), Le Bas and Sierra (2002), Patel and Vega (1999), von Zedtwitz and Gassmann (2002)	Gassmann and Han (2004), Motohashi (2012), Schanz et al. (2011), von Zedtiwitz (2004)	Diminin et al. (2012)
Production driven	Gammeltoft (2006), Gassmann and von Zedtwitz (2002), Le Bas and Sierra (2002), Patel and Vega (1999), von Zedtwitz and Gassmann (2002)	Gassmann and Han (2004), Motohashi (2012), von Zedtiwitz (2004)	
Technology driven (pull)	Gammeltoft (2006), Gassmann and von Zedtwitz (2002), Gerybadze and Reger (1999), Kuemmerles (1999), Le Bas and Sierra (2002), Patel and Vega (1999), von Zedtwitz and Gassmann (2002)	Gassmann and Han (2004), Motohashi (2012), von Zedtiwitz (2004)	Diminin et al. (2012)
Innovation driven (push)	Cantwell and Mudambi (2005), Gammeltoft (2006), Gassmann and von Zedtwitz (2002), Gerybadze and Reger (1999), Kuemmerles (1999), Le Bas and Sierra (2002), Patel and Vega (1999), von Zedtwitz and Gassmann (2002)	Schanz et al. (2011)	Diminin et al. (2012)
Cost driven	Gammeltoft (2006), Gassmann and von Zedtwitz (2002), von Zedtwitz and Gassmann (2002)	Gassmann and Han (2004), Motohashi (2012), von Zedtiwitz (2004)	
Policy driven	Gammeltoft (2006), Gassmann and von Zedtwitz (2002), von Zedtwitz and Gassmann (2002)	Gassmann and Han (2004), Motohashi (2012), von Zedtiwitz (2004)	

2.2 Organisational Structures of Internationalised R&D

Argyres and Silverman (2004) summarise three types of R&D organisational structure in large firms: a centralised structure that one central R&D report directly to the HQ; a decentralised structure that research is conducted exclusively within division or business unit; and a hybrid structure that combines both features. Centralised structure might allow companies to exploit economics of scale, scope and knowledge spillover from research projects, while decentralised structure might allow firms to enjoy efficiency created by improved information processing and reduced scope of managerial opportunism.

Von Zedtwitz (2004) take a step further and try to identify a set of underlying determinants of the degree of decentralisation of R&D projects and put forward four determining attributes of the project involved: (1) type of innovation: Incremental versus radical; (2) nature of the project: Systemic versus autonomous; (3) knowledge mode: Explicit versus tacit; and (4) degree of resource bundling: Redundant versus complementary. They then argue that the project of radical innovation, systemic project work, prevalence of tacit knowledge and the presence of complementary resources requires a more centralised approach, while that of incremental innovation, autonomous project work, prevalence of explicit knowledge, and the presence of redundant resources is compatible with a more decentralised approach.

Based on the location of research and development functions, four types of R&D operational structure are identified by von Zedtwitz and Gassman (2002)—domestic R & domestic D, domestic R & dispersed D, dispersed R & domestic D, and dispersed R & dispersed D. From the view of knowledge management, Birkinshaw (2002) distinguishes research units by the types of knowledge involved in the activities. First is self-contained R&D centres, with knowledge assets that are high on observability and low on mobility; second is modular centres with low observability and high mobility assets; the last one is home-based centres, with knowledge assets low in both observability and mobility. He then suggests that the structure of R&D networks is also related to the knowledge type: (1) loosely coupled network with low knowledge mobility and (2) integrated network with low knowledge observability.

2.3 Practical Issues Related to Overseas R&D Management

Von Zedtwitz and Gassmann (2002) examine the main challenges in establishing an international R&D unit and conducting transnational R&D projects. One significant barrier for companies performing international R&D is the physical distance among R&D units. Compared to local R&D project, international R&D projects need more efforts in communication, coordination and information exchange. The separation of R&D units might cause not-invented-here syndrome and compartmentalisation within the company's R&D functions. It is also difficult to create a coherent working culture, exchange tacit knowledge and build trust in the distanced R&D unit.

In addition to the communication and coordination problems caused by physical distance, several issues have been identified in managing a R&D unit away from homeland. These issues include:

- Managing culture diversity (Gassmann and von Zedtwitz 2002; Gassmann and Han 2004; von Zedtiwitz 2004)
- Recruiting, training and retaining of managers and employees (Gassmann and von Zedtwitz 2002; Selmer 2002; Kim and Oh 2002; Gassmann and Han 2004; von Zedtiwitz 2004)
- Creating synergies with other R&D units (Gassmann and von Zedtwitz 2002; Yang and Jiang 2007)
- Retention, integration and utilisation of dispersed know-how (Birkinshaw 2002; Gassmann and von Zedtwitz 2002; von Zedtiwitz 2004)
- Protecting intellectual property (Quan and Chesbrough 2010; Yang and Jiang 2007; Zhao 2006; Gassmann and Han 2004)
- Government bureaucracy and lack of transparency (Gassmann and Han 2004)

3 Methodology

Due to the explorative nature of this study, we use case analyses of six international engineering practices between Europe and China to illustrate different growth paths adopted by the case companies. Tables 3 and 4 summarise the characteristics of these companies involved in the international practices. The companies are ensured confidentiality in case studies; therefore, they are disguised until formal approval.

Table 3 A summary of overseas R&D units set up by European companies

European company	LifeStyle	Mechanic	BodyGuard
Product market	Consumer electronics	Mechanic components	Electronic products
Annual turnover (US dollars)	30 billion	10 billion	1 billion
% of sales from China in 2013	32% in Asia/Pacific region	24% in Asia/Pacific region	Less than 5%
Headquarter	West European country	Scandinavian country	Scandinavian country
Global R&D network	The USA, Europe and Asia	The USA, Europe and Asia	China only
R&D location in China	Shanghai	Shanghai	Shanghai
Production location in China	13 manufacturing sites in China	Multiple locations	Outsourced to Chinese manufacturers
University collaboration	Top universities in Shanghai and Zhejiang	Top universities in Beijing	None
Year set up in China	2002	2010	2013
Size of the R&D unit	110 employees	40	N/A
Entry mode to China	Greenfield	Greenfield	Greenfield

Table 4 A summary of overseas R&D units set up by Chinese companies

Chinese company	Messenger	Thomas	ShopSafe
Product market	Telecommunication	Transportation	Electronic equipment
Annual turnover (US dollars)	30 billion	1 billion	60 million
% of sales from overseas	65%	N/A, mainly Chinese market	89%
Headquarter	Southern China	Mid-China	East coast China
Global R&D network	The USA, Europe and Asia	Only one	Only one
R&D location in Europe	UK	UK	Italy
Production location in Europe	Eastern European country	UK, belongs to the acquired company	None
University Collaboration	With multiple UK universities	None	None
Year set up in Europe	2010	2008	2011
Size of the R&D unit	300 R&D staffs	40 R&D staffs	Not disclosed
Entry mode to Europe	Acquisition	Acquisition	Acquisition

The cases we selected are all in manufacturing and engineering sectors. This is because these companies are the main source of global R&D activities, and most of the recoded data are available in these sectors. A practical factor worth mentioning is the accessibility to these companies.

We developed a semi-structured data collection protocol focusing on the location choice, R&D network, R&D output and strategic objective of these overseas R&D units. Informants were top management team members, director of research or senior managers. Half of them had personally involved in the initial setting up process. Case companies' annual reports and other public information were also collected to triangulate the information provided by the informants. Table 5 shows a list of informants.

Table 5 A list of informants

Company name	Informants
LifeStyle	Director of China Research Centre, Business Director in research in HQ, Strategy Director in research in HQ and Program Manager
Mechanic	Innovation Manager
BodyGuard	Director of Business
Messenger	General Manager of a foreign branch
Thomas	Based on public available interview data
ShopSafe	President, CTO

4 Cross-Case Analysis

Tables 6 and 7 present the comparison of motives, location choice factors, entry mode, performance and management issues of the six case companies, followed up by a comparative analysis.

When three MNCs from Europe decide to set up R&D facilities in China, the main decision factor is to better access to the Chinese market, by meeting local customers' need. LifeStyle and Mechanic already have a big presence in China. They choose to locate in Shanghai, where they can get support from the business offices already in operation. BodyGuard's sale in China does not have much share in its last financial year, but it realises that China is a very important market for them and they need to have some R&D presence in China.

On the contrary, two Chinese companies Messengers and Thomas go to the Europe specifically for technology, while ShopSafe has a very clear motive to directly introduce acquired product to the Chinese market. All three overseas units are acquired; hence, there is not much decision space for choosing a specific location. The length of acquisition usually lasts less than one year from open to close. The three Chinese companies have to move and decide very quickly when the opportunity window appears. In the case of ShopSafe, the company just went public and had cash for investment. The integration of acquired unit with the parent company did not go so well. The president would rather describe it as a learning process for international acquisition. The company acquired three more companies outside China since then.

Table 6 Comparison across European companies

	LifeStyle	Mechanic	BodyGuard
Main motives	Market driven "LifeStyle at that time has a big factory of optical storage in Shanghai, so the topic was chosen..." - Director of research in China	Market driven "bringing innovation and technical knowledge closer to customers in Asia to better meet local customers' needs" - Company report	Market driven "...we had a little sales in China, but we have to be there to understand customers" - Director of Business
Location choice: Main decision factor	Operational efficiency "...reason for that was the headquarter was in Shanghai and most business were in Shanghai, you like to be close to them.. semi-conductor division has big activity in Shanghai... so that was the reason to chosen to be able to connect to those" - Director of research in China	Support business "There are several candidates... we take into consideration many different factors...close to the plants and customers are the key factors." - Innovation manager	Operational efficiency "...because we already have an office in Shanghai" - Director of Business
Entry mode	Greenfield	Greenfield	Greenfield
Current performance	Number of employee grows from 15 to 100	Just moved to the new R&D building	Just set up
Management issues	Research support from the government "If we applied for participating in Chinese national programmes. We never succeed, none of the internationals succeed. Then we found we were not treated equally" - Director of research in China	Operational cost "At the beginning, we did want to leverage the cheap labour in China, but the reality does not justify this motivation. We starts to measure the cost versus efficiency/productivity and the value created" -Innovation manager	N/A

Table 7 Comparison across Chinese companies

	Messenger	Thomas	ShopSafe
Main motives	Innovation and technology driven "After acquisition we integrated the company's R&D team into Messenger's own research team, Messenger's optic R&D capabilities can be significantly enhanced" -CEO of acquired company "…the UK is at the forefront of developments in wireless, multimedia and advanced communications" - CEO of UK branch	Innovation and technology driven "Particularly they want to invest technologies and facilities we have here, so we would become a leader in technology" - CEO of acquired company	China-market driven "We want a product that can upgrade our market position" - President of ShopSafe
Location choice: Main decision factor	By acquisition, not much location decision involved	By acquisition, not much location decision involved	By acquisition, not much location decision involved
Entry mode	Acquisition, all R&D staffs are kept in the centre	Acquisition, all R&D staffs are kept in the company "The strategy Thomas discussed with us is to retain our operation here" - CEO of acquired company	Acquisition, all staffs are kept in the company "Buying the company is a very quick decision. I see the opportunity and we happen to have a lot of cash." - President

(continued)

Table 7 (continued)

	Messenger	Thomas	ShopSafe
Current performance	Aggressive growth "The acquired centre has 50 R&D staff in 2012 and we want to boost the number to 300 by the end of 2015." - CEO of UK branch	Moderate growth With financial support from Thomas, the acquired company invested a new R&D centre. The R&D personal increased from 12 to 40 in 2013. The acquired company also expanded to new market areas with the help of Thomas	Under expectation "This new product technology and sales channel are very different from our own business, we are struggling to convince retailers to use our products….It is the first foreign company we have acquired…I would think it as a learning process" - President
Management issues	The centre will be mainly dedicated to Messenger's R&D priorities. "We expect all contracted projects to be completed and current customers are being assisted to find alternative source of supply" - CEO of acquired company	High autonomy "They give us a high degree of autonomy, and they did not place a Chinese manager at the top after acquisition"- CEO of acquired company	High autonomy "We acquired the company, but we still work as two separate companies. We have a small unit in China localising the product" - President

It is also worth noting that all three European companies sent the directors of R&D from the HQs. The majority of their research or engineering staffs are Chinese. The three Chinese companies did not change the management team and research staffs after acquisition. Messenger and Thomas continue to invest in R&D, and both acquired units expend significantly after the acquisition. In terms of integration, Messenger already has a comprehensive R&D network, and the acquired R&D centre was immediately given new agenda. Thomas and ShopSafe were kept relatively independent from the parent companies. Both acquired units were given new opportunities to embrace a big market through parent companies, either through technology transfer or through direct product sales.

The three European companies more or less experienced management challenges we have discussed in the previous section. Culture diversity, staff recruiting and retaining, and government relationships are the key challenges for them. Interestingly enough, although China has a weak IP protection regime, LifeStyle and Mechanic both worked out ways to protect their intellectual properties. The three Chinese companies do not seem to have local management issues as the units were given high autonomy in daily operation. The biggest challenge is in creating synergy: How could acquired units fit into parent companies' research portfolio and product portfolio?

5 Discussion and Conclusion

The motivations of investing R&D in China or Europe in our case examples are consistent with the literature. There is a clear pattern that European companies invest R&D in China for market purpose, while Chinese companies go to the Europe for better technology. This is a logical choice. However, two of the European companies we interviewed have the ambition to expend their R&D units to full-scale research centres: "We see China as a second home" quoted by one interviewee. LifeStyle is now doing research for global market, and Mechanic also provides global engineering services. Although there are still concerns about transferring core technology to China, with the growing of China's market and local science and technology capabilities, we expect to see more in-depth R&D activities from Western MNCs.

Chinese companies tend to use acquisition to minimise risks of operating in an unfamiliar environment. Their target is very clear: Complimentary technological resources. With some information asymmetries in acquisitions, it is a trial-and-error process for Chinese companies to embrace the advantages of internationalisation of R&D. With the accumulation of experiences and globalisation of Chinese firms, we would expect to see more Chinese companies operating R&D overseas.

This is only the first step of exploratory research. We did find Chinese companies have a different pattern in conducting global engineering operations. It would be beneficial to both research and practice if we can further explore the challenges and corresponding countermeasures by Chinese MNCs.

References

Argyres, N., & Silverman, B. (2004). R&D, Organization structure, and the development of corporate technological knowledge. *Strategic Management Journal, 25,* 929–958.

Asakawa, K., & Som, A. (2008). Internationalization of R&D in China and India: Conventional wisdom versus reality. *Asia Pacific Journal of Management, 25*(3), 375–394.

Bartlett, C. A., & Ghoshal, S. (1990). Managing innovation in the transnational corporation. In C. A Bartlett, Y. L. Doz & G. Hedlund (Eds.), *Managing the Global Firm.* New York: Routledge.

Birkinshaw, J. (2002). Managing internal R&D networks in global firms what sort of knowledge is involved? *Long Range Planning, 35,* 245–267.

Buckley, P., & Casson, M. (1998). Analyzing foreign market entry strategies: Extending the internalization approach. *Journal of International Business Studies, 29*(3), 539–561.

Cantwell, J., & Mudambi, R. (2005). MNE competence-creating subsidiary mandates. *Strategic Management Journal, 26,* 1109–1128.

Di Minin, A., Zhang, J., & Gammeltoft, P. (2012). Chinese foreign direct investment in R&D in Europe: A new model of R&D internationalization? *European Management Journal, 30*(3), 189–203.

Eisenhardt, K. (1989). Making fast strategic decisions in high-velocity environments. *Academy of Management Journal, 32*(3), 543–576.

Florida, R. (1997). The globalization of R&D: Results of a survey of foreign affiliated R&D laboratories in the USA. *Research Policy, 26*(1), 85–103.

Gammeltoft, P. (2006). Internationalization of R&D: Trends, drivers and managerialchallenges. *International Journal of Technology and Globalization, 2* (1–2), 177–199.

Gassmann, O., & Han, Z. (2004). Motivations and barriers of foreign R&D activities in China. *R&D Management, 34*(4), 423–437.

Gassmann, O., & vonZedtwitz, M. (1999). New concepts and trends in international R&D organization. *Research Policy, 28,* 231–250.

Gassmann, O., & vonZedtwitz, M. (2002). Organization of industrial R&D on a global scale. *R&D Management, 28*(3), 147–161.

Gerybadze, A., & Reger, G. (1999). Globalization of R&D: Recent changes in the management of innovation in transnational corporations. *Research Policy, 28,* 251–274.

Hong, E., & Sun, L. (2006). Dynamic of internationlisation and outward investment: Chinese corporations' strategies. *The China Quaterly, 187,* 611–634.

Johnson, B. (2008). *Development of a global facility location analysis tool.* Cambridge, MA: MIT.

Kim, B., & Oh, H. (2002). Economic compensation compositions preferred by R&D personnel of different R&D types and intrinsic values. *R&D Management, 32*(1), 47–59.

Kuemmerle, W. (1999). Foreign direct investment in industrial research in the pharmaceutical and electronics industries: Results from a survey of multinational firms. *Research Policy, 28*(2), 179–193.

Le Bas, C., & Sierra, C. (2002). Location versus home country advantages in R&D activities: Some further results on multinationals' locational strategies. *Research Policy, 31,* 589–609.

Motohashi, K. (2012). Management of offshore R&D in China : Cross country differences in motivation and performance. *Institutions and Economies, 4*(2), 101–120.

Patel, P., & Vega, M. (1999). Patterns of internationalisation of corporate technology: Location vs. home country advantages. *Research Policy, 28,* 145–155.

Quan, X., & Chesbrough, H. (2010). Hierarchical segmentation of R&D process and intellectual property protection: Evidence from multinational R&D laboratories in China. *IEEE Transactions on Engineering Management, 57*(1), 9–21.

Schanz, C., Husig, S., Dowling, M., & Gerybadze, A. (2011). Low cost–high tech innovations for China: Why setting up a separate R&D unit is not always the best approach. *R&D Management, 41*(3), 307–317.

Selmer, J. (2002). The Chinese connection? Adjustment of western vs. overseas Chinese expatriate managers in China. *Journal of Business Research, 55,* 41–50.

Sun, Y., & Wen, K. (2007). Uncertainties, imitative behaviours and foreign R&D location: Explaining theover-concentration of foreign R&D in Beijing and Shanghai within China. *Asia Pacific Business Review, 13*(3), 405–424.

Sun, Y., Von Zedtwitz, M., & Fred Simon, D. (2007). Globalization of R&D and China: An introduction. *Asia Pacific Business Review, 13*(3), 311–319.

United Nations Conference on Trade and Development (UNCTAD). (2005). UNCTAD Survey on the internationalization of R&D: Current patterns and prospects on the internationalization of R&D. Geneva: UNCTAD.

Von Zedtwitz, M., & Gassmann, O. (2002). Market versus technology drive in R&D internationalization: Four different patterns of managing research and development. *Research Policy, 31,* 569–588.

VonZedtwitz, M. (2004). Managing foreign R&D laboratories in China. *R&D Management, 34*(4), 439–452.

Wldmeir, P. (2012, Nov 13). China offers a taste of R&D to come. *Financial Times.*

Yan, A., & Gray, B. (1994). Bargaining power, management control, and performance in United States-China joint ventures: A comparative case study. *Strategic Management Journal, 37*(6), 1478–1517.

Yang, Q., & Jiang, C. X. (2007). Location advantages and subsidiaries' R&D activities in emerging economies: Exploring the effect of employee mobility. *Asia Pacific Journal of Management, 24*(3), 341–358.

Yin, R. (2003). *Case study research: Design and methods.* Thousand Oaks: SAGE.

Zhao, M. (2006). Conducting R&D in countries with weak intellectual property rights protection. *Management Science, 52*(8), 1185–1199.

Authors' Biography

Quan Zhou is a doctorial researcher at the national institute of innovation management, Zhejiang University. His research focusses on strategic management and innovation management.

Xiaobo Wu is Qiushi Chair Professor of Strategy and Innovation Management, Dean of School of Management, Cheung Kong Scholar Chair Professor of Ministry of Education, China. He is Director of National Institute for Innovation Management, and Director of Global Zhejiang Entrepreneur Research Centre, at Zhejiang University. He is also the director of the Zhejiang University-Cambridge University Joint Research Centre for Global Manufacturing and Innovation Management. And He used to serve as the Senior Vice President of International Council for Small Business.

Yongjiang Shi is research director of the Centre for International Manufacturing (CIM) at the Institute for Manufacturing, Cambridge University. He has undertaken extensive research into international manufacturing network configurations. His research interests include global manufacturing networks, business ecosystem development, the strategies of multinational corporations in China, and sustainable industrial systems.

7

Engineering Value Chain Simulation and Innovation

Tao Zhang and Yufeng Zhang

1 Introduction

Engineering is the art of transforming innovative ideas and technologies into useful products and services, thus creating values to humankind (Zhang and Gregory 2011). Traditionally, large-scale engineering operations tended to be centralised and were funded and conducted by some resource-/knowledge-intensive engineering firms. With the development and integration of the world's economy, particularly the rise of emerging economies that provide rich engineering resources and markets, engineering activities are increasingly dynamic and dispersed on a global scale (NAE 2008). Innovative ideas are generated, transformed and realised across disciplinary, organisational, geographical and even cultural

T. Zhang (✉)
Birmingham University, Birmingham, UK
e-mail: T.Zhang.1@bham.ac.uk

Y. Zhang
Birmingham Business School, University of Birmingham, Birmingham, UK
e-mail: zhangys@bham.ac.uk

Y. Zhang and M. Gregory (eds.), *Value Creation through Engineering Excellence*,
https://doi.org/10.1007/978-3-319-56336-7_7

boundaries (House of Commons 2009; Zhang and Gregory 2011). It is a trend for an engineering firm to establish engineering centres in foreign countries in order to access local engineering resources and markets. Competitions take place globally, and many engineering giants regard internationalisation as a key development strategy.

In addition to this "internationalisation of engineering" trend, another challenge engineering firms are faced with is the fast-changing customer demands in today's dynamic business environment (Zhang et al. 2008). In order to better satisfy and continuously deliver benefits to customers, an engineering firm needs to work closely with customers, using case-specific resources to understand and communicate with each individual customer. This project- or customer-orientation nature of engineering services requires effective and efficient collaboration and coordination among different business units within an engineering firm. Therefore, engineering activities in today's global business environment present a "network" character. Network concepts have been widely adopted to study engineering operations, e.g. coordination mechanisms (De Meyer 1993; Backhouse and Brookes 1996), organisational structures and/or industrial practices (Thamhain 1992; Tripathy and Eppinger 2011; Zhang et al. 2014). A "global engineering networks" (GENs) concept has been proposed (Zhang and Gregory 2011), and theories and practices about the management and operations of GENs have been discussed intensively in recent studies (Zhang et al. 2007, 2008). One key idea of managing GENs is the "engineering value chain" (EVC), which has been proposed based on the synthesis of the value chain literature in different domains, e.g. the traditional value chain concept, innovation value chain, virtual value chain, value networks and product life cycle management (Zhang and Gregory 2011). An EVC identifies the key engineering activities that can create customer value and develop the overall competitiveness of a firm in five stages along a chain. These value-creating engineering activities are interconnected and distributed across different business units and locations. Together, they form a network that can help a firm create value and develop overall competitiveness via achieving effectiveness and efficiency (Zhang and Gregory 2011).

EVCs present a lot of opportunities for researchers with different methodological expertise. So far, the major research methods in EVC studies are case studies (e.g. Zhang et al. 2007, 2008; Zhang and Gregory 2011) and empirical studies (e.g. MacBryde et al. 2013). We argue that the network character of EVC makes simulation methods, especially agent-based simulation, a particularly useful method for studying EVCs. The engineering firms (or business units) with different engineering activities can be modelled as heterogeneous agents along an EVC. Via some collaboration and coordination mechanisms, these agents can achieve value creation. Agent-based simulation models can help to measure the effectiveness and efficiency of these collaboration and coordination mechanisms and demonstrate the dynamic process of value creation. In this chapter, we specifically elaborate on how agent-based simulation is suited to studying EVC, and demonstrate a research case.

This chapter is organised as follows. Section 2 introduces engineering value chain and its characteristics. Section 3 elaborates on agent-based simulation methods. Section 4 presents a research case. Section 5 discusses the usefulness of agent-based simulation for EVC research. Section 6 concludes the chapter.

2 The Engineering Value Chain (EVC)

In a competitive and dynamic business environment, value refers to "the sum a buyer is willing to pay for what a supplier delivers" (Zhang and Gregory 2011, p. 740). From a traditional manufacturing firm's perspective, there are two types of value-adding activities: primary activities and support activities. Primary activities include all business activities that can produce and market physical products, e.g. inbound logistics, manufacturing of goods, outbound logistics, marketing and sales (Porter 1985). Support activities are to support the value creation of primary business activities, including business activities such as procurement, human resource management, technology management and organisational structure (Porter 1985). These primary and support activities constitute a value chain, creating value that a firm can deliver to its customers.

The value chain concept has been further examined beyond the traditional manufacturing setting. Hansen and Birkinshaw (2007) proposed the concept of innovation value chain, which views innovation as an integrated chain of activities that transforms ideas into valuable commercial outputs beyond traditional finished goods and services. Essentially, an innovation value chain contains three stages: idea generation, conversion and diffusion. Along an innovation value chain, six critical tasks are allocated across the three stages: internal sources, cross-unit sourcing, external sourcing, selection, development and company-wide spread of the idea (Hansen and Birkinshaw 2007). These critical value-adding tasks would enable a company to effectively and efficiently turn innovative ideas into commercial outputs that customers are willing to buy.

In today's fast-changing competitive business environment, the fundamental mechanism of value creation is satisfying customers better and quicker than competitors, as customers are generally willing to pay premiums for the best available offerings in the market. Value creation in a market cannot be achieved by one single firm—in order to create value (i.e. satisfying customers better and quicker than competitors), a firm must work with different economic partners, e.g. suppliers, distributors, business partners, strategic alliance, customers. This constellation of actors forms a value-creating system where the firm develops competitive advantages and co-create value with its partners. The strategic task for the firm in the value-creating system, therefore, is reconfiguring its roles and relationships with these partners to "mobilise the creation of value in new forms and by new players" (Zhang and Gregory 2011, p. 740) in response to the fast changes in the dynamic business environment.

The concept of engineering value chain (EVC) is grounded on traditional value chain theories (especially the innovation value chain). An EVC focuses on the transformation process from innovative ideas into valuable products/services. For engineering firms, this process is often intangible, requires the expertise and experienced engineers, and involves coordination and collaboration between the firm and its partners in a value-creating system (Normann and Ramirez 1993; Gereffi et al. 2005). Figure 1 presents an overview of the EVC with illustrative examples adapted from Zhang and Gregory (2011).

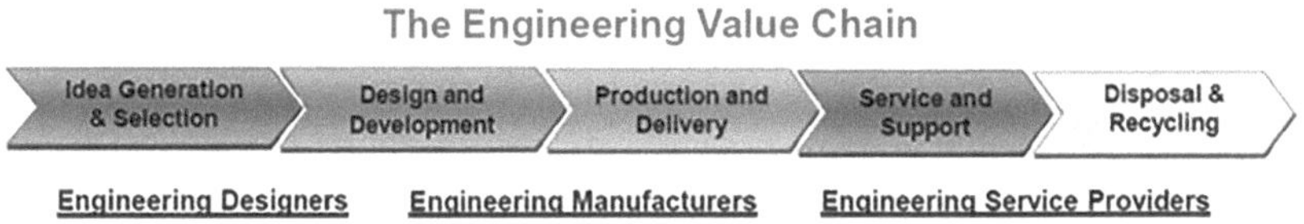

Fig. 1 Engineering value chain overview. adapted from Zhang and Gregory (2011)

Essentially, an EVC is composed of five categories of engineering activities that can create customer value and thus the competitive advantages of a firm: (1) idea generation and selection, (2) design and development, (3) production and delivery, (4) service and support and (5) recycling and disposal. These five categories of engineering activities are interconnected. They achieve value creation and develop the competitive advantages of an engineering firm via improving the "performance of engineering operation from the perspectives of efficiency and effectiveness" (Zhang and Gregory 2011, p. 743). Effectiveness reflects how closely the outputs of an engineering firm meet customers' needs, i.e. how much better and quicker the firm can satisfy customers' needs. An EVC can achieve effectiveness through, e.g. quick response environmental changes, product innovation, mobile engineering resources and flexible operation approaches. Efficiency reflects how economically resources are utilised to produce the outputs. It can be achieved through, e.g. international operation synergies, resource sharing, economies of scale, etc.

An EVC presents a network character. The activities along an EVC belong to different business units which are interconnected. These networked business units can be configured to achieve efficiency and effectiveness. Basically, there are three types of EVC network configuration: efficiency-oriented, innovation-oriented and flexibility-oriented (Zhang and Gregory 2011). These configuration mechanisms would enable an EVC to create customer value from different perspectives. Thus, it is of academic significance to study these configuration mechanisms. Among different methodological approaches to study EVCs, agent-based simulation presents some unique advantages and is particularly useful.

3 Agent-Based Simulation

Theoretically speaking, agents are the constituent units of a multi-agent system (MAS); agents are autonomous, behave on their own and interact with each other in an MAS (Russell and Norvig 2003). The behaviour and interactions of agents produce the global behaviour of an MAS. However, this type of global behaviour of an MAS cannot be traced back to the behaviour and interactions of its constituent agents. Agents can be objects that are of low- or medium-level intelligence (e.g. machines and software programmes, which just have some functional, procedural or algorithmic search, find and processing approach), or objects that are highly intelligent (e.g. human beings and societies) (Russell and Norvig 2003). Agent-based simulation is a computational modelling approach to study MASs. An agent-based model is composed of individual agents, commonly implemented in software as objects. Agent objects have states and rules of behaviour. Running an agent-based model simply amounts to instantiating an agent population, letting the agents behave and interact, and observing what happens globally (Axtell 2000) (See Fig. 2). Thus, a unique advantage of agent-based simulation is that almost all behavioural attributes of agents can be captured and modelled. Agent-based simulation is widely adopted in studying MASs, particularly those with intelligent human beings (e.g. markets, societies and organisations; for further information about agent-based simulation and its applications, please see Galán et al. (2009), Farmer and Foley (2009).

Using ABS to study EVCs requires us to treat the business units along an EVC as autonomous, independent and intelligent agents of different value-creating activities. An agent's particular behaviour is decided by its value-creating activities on the EVC. The links between these agents are decided by their relationships along the EVC. These agents, physically distributed in different geographic areas, constitute the networked EVC through some coordination and cooperation mechanisms. Agent-based models can enable us to understand how these mechanisms can lead to the system-level efficiency and effectiveness of the EVC in a dynamic manner. Figure 3 demonstrates the prototype of an agent-based model of EVC. The prototype shows that there are a lot of opportunities for

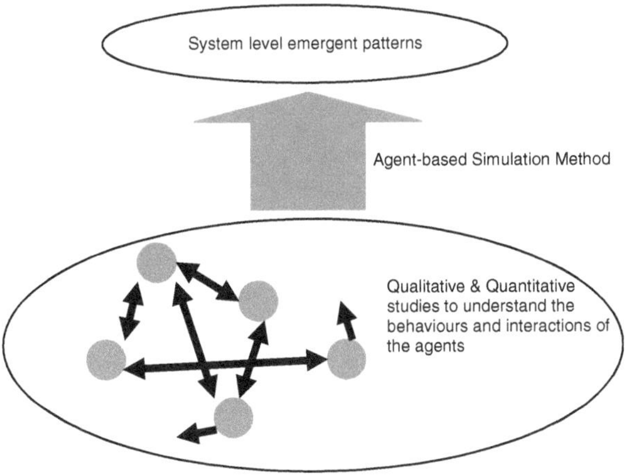

Fig. 2 Agent-based simulation

agent-based simulation in studying EVCs, e.g. designing and testing value-creating mechanisms, optimising the relationship of an EVC with other EVCs (i.e. inter-firm relationships), modelling the market (i.e. firm-customer relationship), innovation diffusion, etc.

4 Modelling Innovation Diffusion: A Case of Effectiveness-Oriented EVC Network Configuration

We demonstrate a case of agent-based simulation of EVC. In this case we use agent-based simulation to study innovation diffusion. Specifically, we look at how different policies/interventions can influence post-adoption user learning using the council-led smart meter deployment in Leeds, UK. This case shows the relationship between an engineering firm and its customers can be moderated by different policies/interventions. Using effective policies we can maximise the benefits of technologies for consumers, thus creating value.

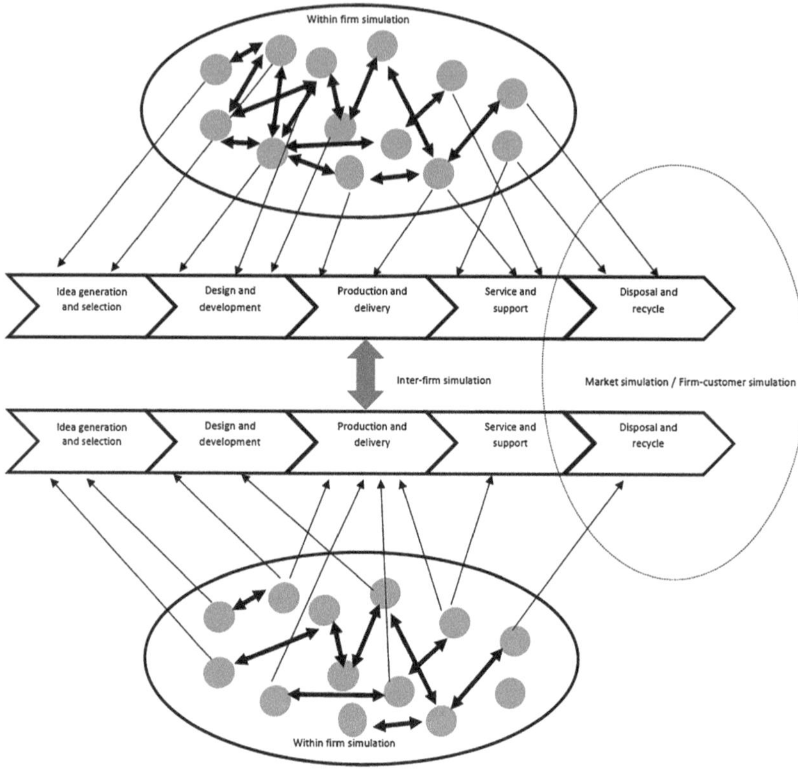

Fig. 3 An agent-based model of EVC

4.1 Background of the Case

As climate change has become a very important global issue, the UK government has set a national target: cutting CO_2 emission by 34% of 1990 levels by 2020. In the UK there are nearly 70 cities and it is increasingly being suggested that each city will have an important role to play for the country's future sustainability (Keirstead and Schulz 2010; Bale et al. 2012; Zhang et al. 2012). With a population of 787,700 (Office for National Statistics 2011), Leeds City Council (Leeds CC) is the second largest metropolitan council in England and also the UK's largest centre for business, legal and financial services outside London.

Having realised its important role for the future of UK sustainability, Leeds CC made a clear statement to voluntarily undertake a target of cutting its CO_2 emission by 40% by 2020. However, like many other UK city councils, Leeds CC has no explicit local energy policies/interventions for achieving this target so far. Currently, Leeds CC is working with the Universities of Leeds and Nottingham and the research council (EPSRC) on a *City Energy Future* project that aims to seek knowledge, experience and develop decision support tools to aid local energy intervention design. Several local energy interventions have been proposed, including setting up a city-level strategic council, developing local district heating networks, running energy-saving education campaigns and deploying smart meters.

Smart metering is a popular area in energy policy research. There is a nuanced debate over the benefits/drawbacks of this technology. On the positive side, smart metering has been empirically proved to be an effective means to influence energy users' behaviour and reduce energy consumption (e.g. Haney et al. 2009; Martiskainen and Ellis 2009; DECC 2012; Xu et al. 2015). On the negative side, there are critiques that smart meters are just cost-saving devices for utility companies and can actually make energy consumers use more energy (Sharma and Saini 2015). In the UK there are also campaigns against the roll-out of smart meters (e.g. Stop smart meters). Clearly, smart metering as an innovative technology cannot benefit the society without energy consumers' behaviour change (Owens and Drifill 2008). How much behavioural change smart meters can lead to is a concern when deploying smart meters has been considered to be an energy intervention (Martiskainen and Ellis 2009). Smart metering intervention has been proposed for Leeds CC because in the UK context there is empirical evidence that smart metering can lead to effective energy consumers' behaviour change and substantial energy saving (Haney et al. 2009; Martiskainen and Ellis 2009). Additionally, smart metering can also help energy consumers in the UK alleviate fuel poverty[1] (Barnicoat and Danson 2015). Many residents living in council-owned properties in Leeds are considered to be in fuel poverty. As Leeds CC has direct control over the council-owned properties, authoritatively installing smart meters in the council-owned properties would potentially be an effective way to help the occupants get out of fuel poverty (through changing their behaviour of using energy, thus reducing their energy consumption and cutting their energy bills).

4.2 Modelling Rationale

The study targets the case of authoritative smart metering adoption in Leeds by developing a computational model using agent-based modelling and simulation (ABMS). ABS is a computational modelling approach to study multi-agent systems (MASs). An agent-based model (ABM) is composed of individual agents, commonly implemented in software as objects. Agent objects have states and rules of behaviour. Furthermore, they often have a memory and the capability to evolve over time. This modelling approach lends itself particularly well for modelling people and their behaviour (Siebers and Aickelin 2011). Running an agent-based model simply amounts to instantiating an agent population, letting the agents behave and interact, and observing what happens globally. Thus, a unique advantage of ABS is that almost all behavioural attributes of agents can be captured and modelled. ABS is widely adopted in studying MASs, particularly those with intelligent human beings (e.g. markets, societies and organisations). In this particular case, the energy users behave and interact in a community, which is a complex social system that is well suited to agent-based simulation.

4.3 Model Design

We draw on the idea of the consumat[2] approach (Janssen and Jager 1999; Jager 2000) and model residential energy consumers (REC) as intelligent agents. Instead of simply adopting the consumat approach, we modify it by considering the empirical evidence from the electricity market. A key difference between the original consumat approach and our modified version is that the original consumat is developed based on a combination of some well-established psychological theories such as human needs, motivational processes, social comparison theory, theory of reason action, etc., while our modified version is developed based on

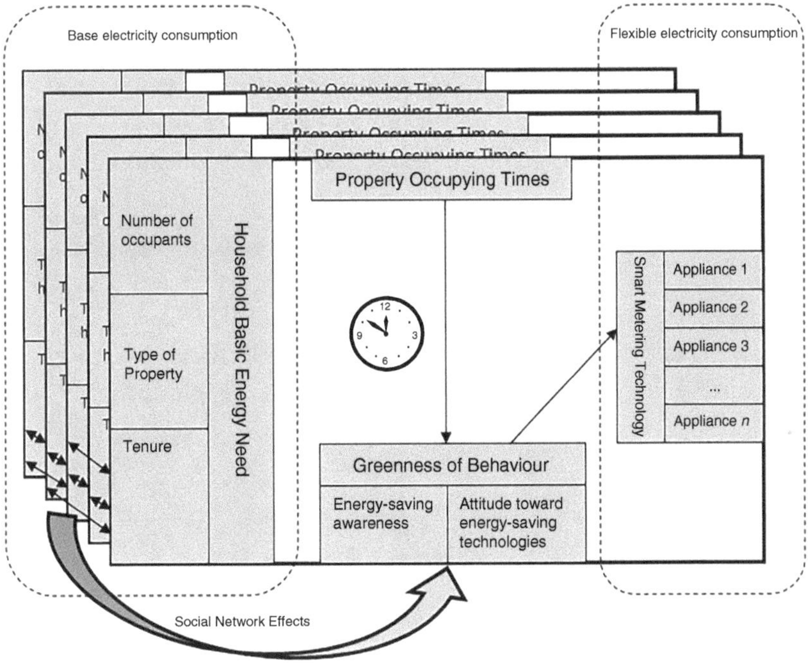

Fig. 4 The overall view of the model

an empirical survey of the residential electricity consumers in the real electricity market.

A template of our REC agents and an overview of the model are shown in Fig. 4. According to our previous research (Zhang et al. 2011, 2012), we consider a household's base electricity consumption which is determined by the household basic energy need, and flexible electricity consumption which is determined by the greenness of the occupants' behaviour and the number and types of electrical appliances in the house. The household basic energy need is influenced by the number of occupants, type of property and tenure. The greenness of the occupants' behaviour is determined by their energy-saving awareness and their attitude towards using energy-saving technologies. The greener the occupants' behaviour is, the higher the probabilities that they respond to the information provided by smart meters (i.e. switching off the unnecessary appliances when the electricity price is high). The REC agents also interact with each other. Their interactions can change their energy-saving awareness and attitude. Property-occupying times determine whether the occupants' green behaviour is effective or not. For example, if the occupants are at home, they can switch on/off the appliances; when they are out of home they cannot do this.

4.3.1 Behaviour of Residential Energy Consumer Agents

We use a state chart to represent the behaviour of REC agents, as shown in Fig. 5. Each REC agent has a set of home electrical appliances. We have carried out a survey to get empirical data about people's social

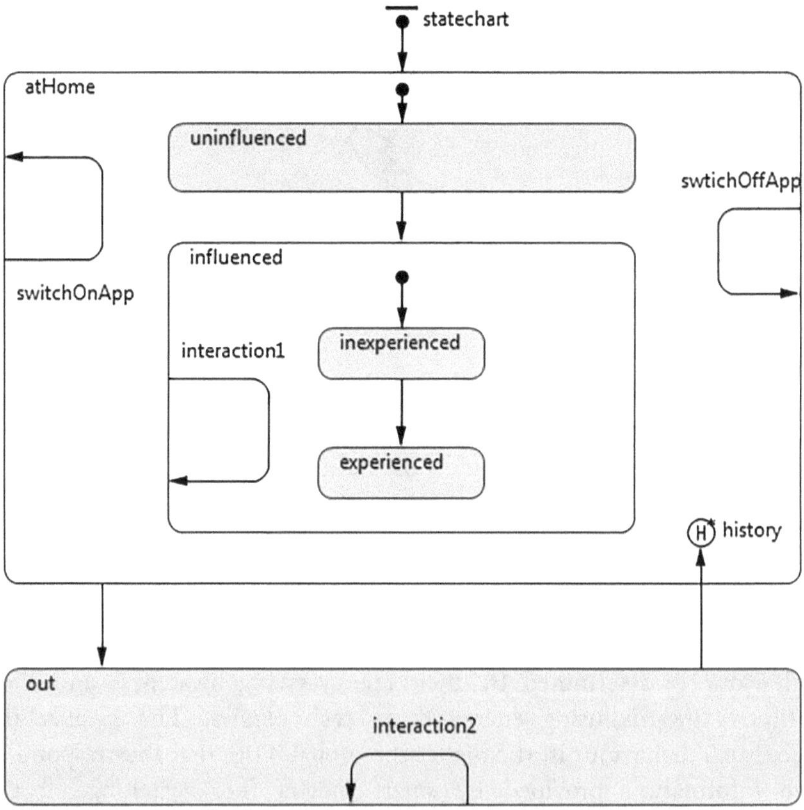

Fig. 5 The state chart of REC agents

Table 1 UK residential energy consumer archetypes (Zhang et al. 2012)

Archetype	**Attributes**		
	Property energy efficiency level	**Greenness of behaviour**	**Duration of daytime occupancy**
1: Pioneer Greens	High	High	Short
2: Follower Greens	Low	High	Short
3: Concerned Greens	Low	High	Long
4: Home-Stayers	High	High	Long
5: Unconscientious Wasters	High	Low	Short
6: Regular Wasters	Low	Low	Short
7: Daytime Wasters	High	Low	Long
8: Disengaged Wasters	Low	Low	Long

demographic attributes, number and types of electrical appliances, levels of awareness about energy, and lifestyles in the Leeds area. Based on the empirical data, we have developed some energy consumer archetypes (Zhang et al. 2012) shown in Table 1. These REC agent archetypes determine the initial parameter settings in the simulation.

In the model we simulate the daily life of energy consumers. An REC agent has two main states: "atHome" and "out". The transition from "atHome" state to "out" state is controlled by the parameter "timeLeaveHome", and the transition from "out" state to "atHome" state is controlled by the parameter "timeBackHome". Both parameters are defined by the archetype of the REC agent. For example, a REC agent of "short daytime occupancy" (e.g. a regular work professional) often leaves home for work at a time between 8:30 a.m. and 9:30 a.m, and goes back home at a time between 5:30 p.m. and 6:30 p.m. These two transitions reflect the "Property-Occupying Times" in Fig. 4.

When the REC agent is in the "atHome" state, it initially is in an "uninfluenced" sub-state if the city council does not implement the smart metering intervention. Otherwise, the REC agent transits from the "uninfluenced" sub-state to the "influence" sub-state. When an REC agent is influenced by the city council's smart metering intervention, it initially is in the "inexperienced" sub-sub-state. Then, the REC agent starts the learning process by trying to use the smart meter. Drawing on the idea of the learning theory, in the simulation model we re-interpret the learning function. We interpret $\boldsymbol{P_t}$ as an REC agent's probability of responding to the information provided by the installed smart meter; $\boldsymbol{M}$ as the REC agent's attitude ($\boldsymbol{A}$) towards smart metering technology; $\boldsymbol{K}$ as the REC agent's energy-saving awareness (ESA); and $\boldsymbol{t}$ as the number of tries. Thus, the REC agent's learning function can be defined as:

$$P_t = A * (1 - e^{-(ESA)*t}) \tag{1}$$

From Eq. 1 we can see that the more an REC tries the technology, the higher the probability that it responds to the information provided by a smart meter. We posit that an REC agent's experience in using smart metering technology is reflected by the probability that it responds to the information provided by a smart meter. The transition from the "inexperienced" sub-sub-state to the "experienced" sub-sub-state is a probability threshold ($\boldsymbol{P_{th}}$), which in the simulation model is initially arbitrarily set at 0.85. The probability ($\boldsymbol{P_t}$) is calculated according to the learning function (Eq. 1). If $\boldsymbol{P_t}$ is larger or equal to $\boldsymbol{P_{th}}$, the REC agent makes the transition from the "inexperienced" sub-sub-state to the "experienced" sub-sub-state. The value of $\boldsymbol{P_{th}}$ reflects the level of experience energy consumers need to gain when they make the best use of smart meters in the real world. The smaller the value of $\boldsymbol{P_{th}}$ is, the easier the energy consumers transit from the "inexperienced" sub-sub-state to the "experienced" sub-sub-state, thus achieving energy efficiency. We also set the probability threshold ($\boldsymbol{P_{th}}$) at 0.8 and 0.9 levels to check the sensitivity of the simulation results.

When REC agent is at home, it can switch on/off appliances. This behaviour is reflected by the two internal transitions "switchOffApp" and "switchOnApp". The initial probabilities for them to occur are determined by its archetype. Once the REC agent has a smart meter, the probabilities for them to occur are determined by $\boldsymbol{P_t}$ calculated by the learning function, i.e. if the REC agent responds to the information provided by the smart meter, it changes its energy use behaviour accordingly.

When an REC agent is influenced by the smart metering energy intervention, it can interact with other REC agents through a network to exchange the knowledge and experience. This is reflected in the state chart by the two internal transitions "interaction1" and interaction2". These two internal transitions can influence the REC agent's attitude (***A***) and energy-saving awareness (ESA) positively or negatively, as shown by the social network effects in Fig. 4. This reflects a reality that in the real energy market sharing knowledge and experience via communications can change consumers' attitude and energy-saving awareness (Owens and Driffill 2008). In the model, we have chosen a small world social network to model the communication channels between REC agents.

4.3.2 Model Implementation

We use AnyLogic 6.7.1 to develop the agent-based model. We use the real map of the Leeds Metropolitan District Area as a background to design the model, as shown in Fig. 6. The model has been implemented on a standard PC with Windows XP SP3.

We develop 1000 REC agents to simulate 1000 residential energy consumers living in council-owned properties. According to our survey, residential energy consumers living in council-owned properties are of archetypes "follower greens", "concerned greens", "regular wasters" and

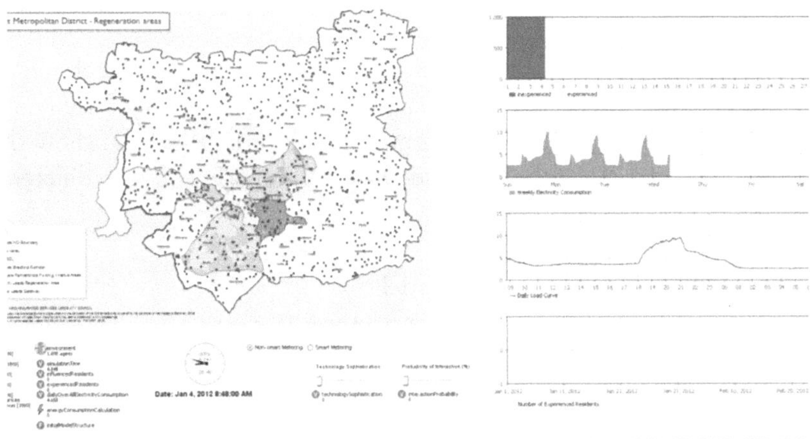

Fig. 6 Interface of the model

"disengaged wasters". Based on the survey statistics, we set "follower greens" 11%, "concerned greens" 13%, "regular wasters" 47%, and "disengaged wasters" 29%.

We set each time step in the simulation model as one minute and simulate the daily life of residential energy consumers and observe and analyse how their learning can result in the dynamics of electricity consumption at the community level. An overview of the model is shown in Fig. 6, where the blue dots are REC agents randomly distributed on the map. When the simulation begins, the REC agents are initial blue. When we choose "smart metering", they turn to red, which means they have been affected by the city council's smart metering energy policy, i.e. the city council has installed smart meters into these council-owned properties. Then, the REC agents start the learning process defined by the learning function (Eq. 1). They also start to interact, and the interactions will influence their learning. With their learning over time, some REC agents turn from red to yellow, which means they transit from "inexperienced" to "experienced" in using smart meters. The parameter settings in the simulation are listed in Table 2. In order to enable other researchers to use the model, we have uploaded the model online at http://www.runthemodel.com/models/879/.

4.4 Experimentation

In order to study the effects of user learning in authoritative technology adoption, we carried out four experiments. Drawing on Rixen and Weigand (2014), we use key performance indicators (KPIs) to show the experiment results. As the purposes of these experiments are different, they have different KPIs.

Experiment 1: Validating the model

The first experiment is meant for model validation. The KPI we use in the experiment is the average daily electricity load curve (kW). We run the model 50 times and gained the average daily load curve (half hourly) of the whole virtual community (1000 REC agents in total) during

Table 2 Simulation parameter settings

Agent archetype	Percentage	*AA*	ESA	TimeLeaveHome	TimeBackHome	P_{th}
Follower greens	11	Normal distribution ($\mu = 0.71$, $\sigma = 0.052$)	Normal distribution ($\mu = 0.74, \sigma = 0.041$)	6:00 to 9:00, random uniform distribution	15:00 to 18:00, random uniform distribution	0.85
Concerned greens	13	Normal distribution ($\mu = 0.69$, $\sigma = 0.050$)	Normal distribution ($\mu = 0.72$, $\sigma = 0.043$)	9:00 to 18:00, random uniform distribution	timeLeaveHome + random (1, 180 min)	0.85
Regular wasters	47	Normal distribution ($\mu = 0.39$, $\sigma = 0.061$)	Normal distribution ($\mu = 0.41$, $\sigma = 0.033$)	6:00 to 9:00, random uniform distribution	6:00 to 9:00, random uniform distribution	0.85
Disengaged wasters	29	Normal distribution ($\mu = 0.22$, $\sigma = 0.037$)	Normal distribution ($\mu = 0.25$, $\sigma = 0.057$)	9:00 to 18:00, random uniform distribution	timeLeaveHome + random (1, 180 min)	0.85

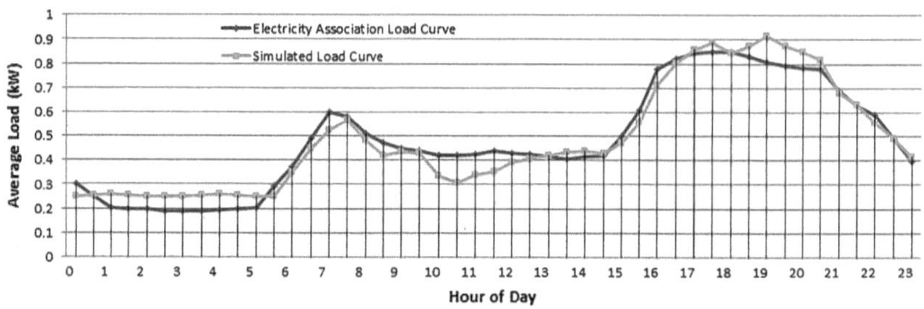

Fig. 7 Simulated load curve versus real load curve

winter time. We then calculate the average daily load curve of an individual REC agent and compare the result with the real standard domestic load curve provided by the Electricity Association (Abu-Sharkh et al. 2006). The comparison in Fig. 7 shows that the two patterns of domestic load curves are quite similar to each other and improve our confidence in the validity of the simulation model.

Experiment 2: Agent transition from "inexperienced" to "experienced"

In the second experiment, we simulated the REC agents' transition from the "inexperienced" to the "experienced" sub-sub-state. The purpose of the experiment is to show the effect of user learning on the community's energy consumption. In this experiment we again use the average daily electricity load curve (kW) as the KPI. We run the model and gain domestic average load curves in both the "inexperienced" scenario and the "experienced" scenario. In the "experienced" scenario, we set the probability threshold ($\boldsymbol{P_{th}}$) at 0.8, 0.85 and 0.9 levels to check the sensitivity of the simulation results, as shown in Fig. 8. In the "inexperienced" scenario all the REC agents are inexperienced, while in the "experienced" scenario 80% of the REC agents have transited from "inexperienced" to "experienced" in using smart meters. Each curve is the average result of 50 runs of the simulation model. The comparison in Fig. 8 shows that the REC agent's transition from "inexperienced" to "experienced" can cause substantial reduction of energy consumption at peak times. This simulation finding has been evidenced by empirical observations in the real electricity

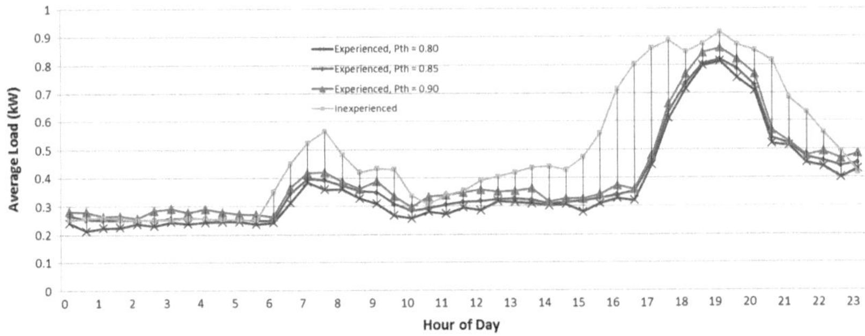

Fig. 8 "Inexperienced" versus "experienced"

market (Haney et al. 2009). The reductions of energy consumption at different P_{th} levels are different. When the P_{th} is at 0.8, 0.85 and 0.9 levels, the average daily energy reduction is 19.5, 18.9 and 18.3%, respectively. This simulation finding reflects that in the real world, if energy consumers can more easily become experienced in using smart meters they tend to become energy-efficient quicker and achieve more energy saving. The simulation finding also means that facilitating user learning can enable users to make better use of an innovative technology, thus gaining more benefit from the technology.

Experiment 3: Learning facilitation

In the third experiment, we study the social network effects on user learning. The purpose of the experiment is to explore the strategies that can facilitate user learning. We use the number of experienced users as the KPI. We adopt a small world (a built-in social network in AnyLogic 6.7.1) and set a contact rate for the REC agents. The contact rate is a probability with which an REC agent contact other REC agents who are connected with it in the social network. When an REC agent contacts other REC agents, it sends a message to them. The message can change other REC agents' energy-saving awareness (ESA) and attitude (***A***). We set the contact rate 0.5, 0.3 and 0.1, and plot the transition trends over 3 month (90 days virtual time) in Fig. 9 (each curve is the average result of 50 runs). From Fig. 9 we can see that contact rate has significant influence on the transition. The higher the contact rate is, the faster the

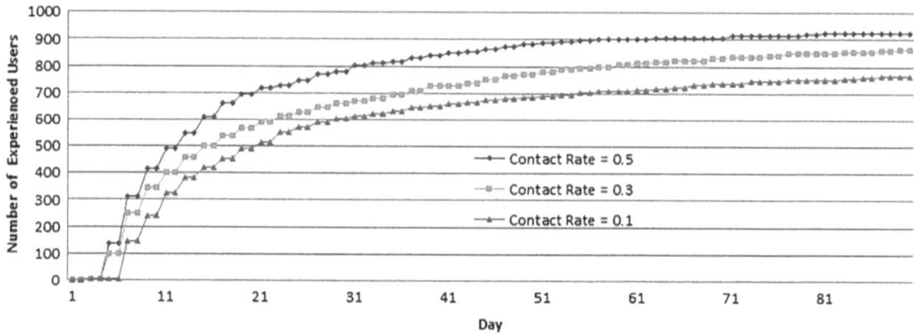

Fig. 9 The influence of contact rate on transition

transition is and the more the final experienced users there are. The simulation finding means that in the real world when an authoritative technology adoption happens, encouraging users to share experience and exchange knowledge about the technology will facilitate user learning.

Experiment 4: Technology discontinuance

In the fourth experiment, we focus on the REC agents' continuation of using smart meters. If all the REC agents can quickly learn how to use smart meters effectively and maintain their interest in using the technology, they can maximise the benefits of the technology. Technology continuance is a very interesting topic in information systems and marketing studies. There are some empirical studies focusing on identifying the factors that can influence users' continuation of using innovative technologies (e.g. Spiller et al. 2007; Parthasarathy and Bhattacherjee 1998; Karahanna et al. 1999). In this simulation study, according to the learning function (Eq. 1) an REC agent's probability of responding to the information provided by smart meters (***P***) is influenced by its attitude (***A***) and energy-saving awareness (ESA). Because of the social network effects, both ***A*** and ESA are changing over time. By exploring how the ***A*** and ESA of discontinuers evolve, we can find out how significant they are in technology discontinuance.

We stick to the transition probability threshold (P_{th} = 0.85). We assume that after an REC agent transits from "inexperienced" to "experienced" sub-sub-state ($P >= P_{th}$), if its ***P*** later drops below P_{th}, it loses its interest in using its smart meter and becomes a discontinuer (however, it will not transit back from "experienced" to "inexperienced" as in the real word a user can lose its interest in using a technology although he/she is still experienced in using it).

Using the ***A*** and ESA as the KPIs, we run the model 50 times (90 days virtual time) and each time we can identify a small proportion (8%) of discontinuers. We collect the statistics of both continuers and discontinuers' ***A*** and ESA and calculate the average of an individual, as shown in Figs. 10 and 11. From a comparison between the two figures we can see that while both continuers and discontinuers have the same trend in their ESA, a key difference between them is their ***A***. Discontinuers drop their ***A***s significantly. The simulation finding shows that attitude is one

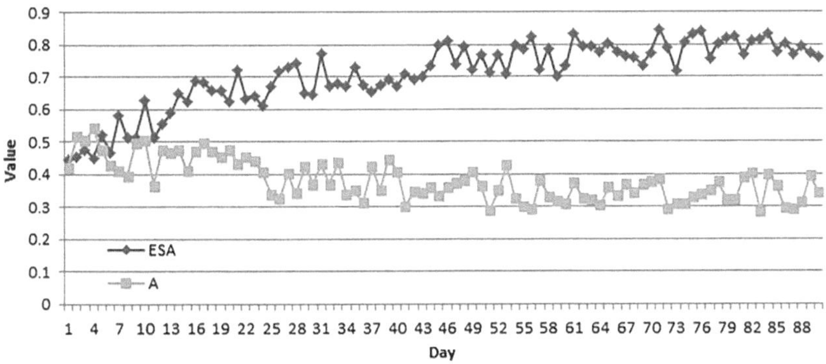

Fig. 10 The average values of individual discontinuers' ***A*** and ESA

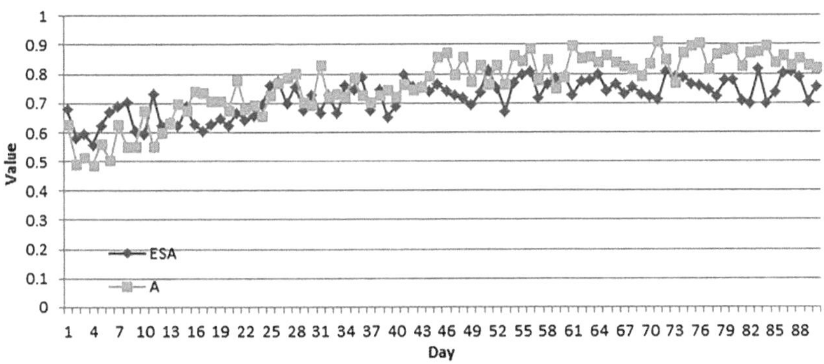

Fig. 11 The average values of individual continuers' ***A*** and ESA

significant factor influencing users' continuation in using a technology. A major reason for users to discontinue to use a technology is that they change their attitude negatively (i.e. decreasing ***A***s). This simulation finding is also in line with some empirical research in information systems studies (e.g. Karahanna et al. 1999).

4.5 Advantages of the Case Study

This study—for the first time—develops an agent-based model to study the effects of user learning in authoritative technology adoption, an area that has been ignored in most other studies. Authoritative technology adoption is quite common, particularly when the authorities systematically upgrade utility infrastructures in communities. Thus, studying authoritative technology adoption can help authority decision-makers make better strategies for maximising the benefits of adopted technologies. In this particular study, we used the case of smart meter deployment in the UK City of Leeds. A unique advantage for using this particular case is that the benefit of the technology (i.e. smart metering) can be quantitatively presented by the reduction of electricity consumption, which we can easily collect with the simulation model (e.g. the energy savings at different levels in Experiment 2).

Our choice of the agent-based simulation method enables us to capture the learning behaviour of users at microlevel well. The REC agents were created based on the residential energy consumer archetypes developed from our empirical survey in the City of Leeds. We have also based the REC agents' learning behaviour on the behavioural learning theory and the learning function which were extensively justified by empirical studies (e.g. Bennett and Mandell 1969; Schiffman et al. 2008). Thus, the REC agents have a solid empirical grounding. We also validated our simulation output against the real-world observation (i.e. real domestic load profile from the Electricity Association). We believe the research outputs are valid and reliable and can provide meaningful decision implications.

4.6 Energy Policy/Intervention Implications

In 2009 the UK Government announced a policy of installing smart meters for all residential energy consumers in GB by 2020. The latest statistical release shows that by 30 September 2014 a total number of 621,600 domestic smart meters were installed, and around 543,900 smart meters are now operating in smart mode in domestic properties in GB. It is expected that city councils play significant roles in the roll-out process (especially the installations of smart meters in council-owned properties). From the study we can see that when authoritative installations of smart meters take place in council-owned properties, enabling energy users quickly to gain experience can be an effective means to reap the benefit of smart metering. Energy users' quicker and easier transitions from the "inexperienced" sub-sub-state to the "experienced" sub-sub-state (i.e. reflected by a smaller P_{th} value) can result in substantial energy saving. This learning process can be facilitated by encouraging contacts via various informational means. For example, city councils can run energy educational programmes, events or energy-saving campaigns, etc., for council house residents, which can encourage their contacts for sharing knowledge and experience about smart metering and energy saving. Moreover, after the installations it is significantly important to develop and maintain energy users' positive attitude towards smart meters so as to encourage them to use the technology continuously. This can be achieved through, for example, using council leaflets to highlight the energy savings and environmental benefits they gain from smart metering.

4.7 Limitations of the Case Study

Although the REC agents developed in the research have a strong empirical root, they cannot perfectly replicate the residential energy consumers in the real world. Thus, it is important to acknowledge the limitations of the research. A first limitation is related to the critique of the behavioural learning theory which sees users as mindless passive objects. Some people believe that although actual users passively accept a technology in authoritative technology adoption, they still need to actively learn how to use it. The current simulation model in the research cannot capture the users' activeness. A second limitation is that in the simulation model the REC agents' archetypes are fixed. In other words, in the simulation there is no way for the REC agents to switch their archetypes. But we note that in the real world the change of archetypes can happen (for example, if a residential energy consumer loses his/her job, he/she might spend more time at home), although the probability for its occurrence is low. A third limitation is related to the critique of using simulation models as decision support tools to guide practice. Numbers and graphs from simulations may impress local authorities and the public, which can help to boost their receptivity to further dialogues and collaboration. However, this is just the beginning. The true utility of the method will only be proven in practice.

5 Discussion

The case demonstrates the usefulness of agent-based simulation for innovation diffusion in EVC studies. We discuss its advantages, opportunities and challenges when using this method in an EVC setting.

Advantages: Compared with traditional top-down empirical methods, agent-based simulation takes a system view using a bottom-up approach, i.e. it models the business units as intelligent agents and examines their value-creating activities and interactions along an EVC. These individual-level activities and interactions can give raise to the whole EVC's system-level efficiency and effectiveness. The value creation process can be dynamically demonstrated via computational simulation. Thus, compared with traditional static empirical methods, agent-based simulation can dynamically analyse the value creation process of an EVC. Value creation mechanisms, i.e. coordination and cooperation mechanisms of the agents, can be designed and implemented in an agent-based model. Their impacts on the efficiency and effectiveness of the value creation of an EVC can be analysed. The outputs of the simulation model can give us implications for the EVC network configuration in the real world. Additionally, traditional empirical methods require large amounts of costly quantitative/qualitative empirical data which may entail ethical concerns; while the parameterisation of an agent-based model primarily relies on secondary data and/or researchers' experience, which can substantially reduce the cost for primary empirical data collection.

Opportunities: An EVC in essence is huge, thus presenting many opportunities for agent-based simulation. At a high abstract level, the whole EVC can be modelled with one agent-based model (like the prototype model). Business units along the EVC are modelled as agents, and coordination and cooperation mechanisms regarding how to improve the whole EVC's efficiency and effectiveness can be analysed and tested. Strategic implications on different types of EVC network configuration, e.g. efficiency-oriented, innovation-oriented and flexibility-oriented can be obtained. At a medium abstract level, the relationship between some business units within an EVC can be modelled. Specific value-creating activities and inter-department/division cooperation and operations mechanisms can be analysed in detail. At a low abstract level, the relationship between an EVC and customers, i.e. the market, can be modelled. Different policies that can govern the market and maximise customer value can be studied. Agent-based simulation has a broad range of applications in EVC studies.

Challenges: As a computational simulation method, agent-based simulation faces some challenges. One general challenge is its questionable validity and reliability. Many researchers criticise agent-based models for being toy models whose outputs lack robustness because many of the parameters in the simulation models are set based on researchers' experience. A specific challenge to the applications of agent-based simulation in EVC research is the lack of understanding of value-creating behaviour of the business units alone an EVC. This is because EVC itself is a new concept that requires more new insightful studies, although we have already made some significant advancement.

6 Conclusion

In this chapter we give our opinion about the usefulness of agent-based simulation in EVC research. We introduce EVC, agent-based simulation and theoretically argue that, because of the network character of EVCs, agent-based simulation is particularly suitable for studying them. We present an agent-based model of innovation diffusion as a case studying effectiveness-oriented innovation diffusion in EVC research. This case shows how policies/interventions can improve/optimise customer value. We then discuss the advantages, opportunities and challenges of using agent-based simulation in EVC studies. We conclude that, although the application of agent-based simulation in EVC studies is still in its infancy and faces a lot of challenges, it is a promising method and will advance in the future with the growing body of empirical EVC research.

Notes

1. A household is said to be in fuel poverty if it needs to spend more than 10% of its income on fuel to maintain a satisfactory heating regime (usually 21 °C for the main living area, and 18 °C for other occupied rooms) (UK National Statistics 2011).
2. The "consumat approach" is a model of human behaviour with a particular focus on consumer behaviour. It is based on concepts and theories from psychology, economics and computer science. The conceptual framework is a kind of meta-model of the many theories in psychology. The computational version of the consumat approach is based on multi-agent simulation. For details, please see Janssen and Jager (1999) and Jager (2000).

References

Abu-Sharkh, S., Arnold, R. J., Kohler, J., Li., R., Markvart, T., Ross, J. N., Steemers, K., Wilson, P., & Yao, R., Kazmerski, Lawrence L. (2006). Can microgrids make a major contribution to UK energy supply? *Renewable and Sustainable Energy Reviews, 10,* 78–127.

Axtell, R. (2000). Why agents? On the varied motivations for agent computing in the social sciences. In C.M. Macal & D. Sallach (Eds.), *Proceedings of the workshop on agent simulation: Applications, models, and tools* (pp. 3–24). Argonne, IL: Argonne National Laboratory.

Backhouse, C., & Brookes, N. (1996). *Concurrent engineering: What's working where*. Hampshire: Gower/Design Council.

Bale, C., Foxon, T., Hannon, M., & Gale, W. (2012). Strategic energy planning within local authorities in the UK: A study of the city of Leeds. *Energy Policy, 48,* 242–251.

Barnicoat, G., & Danson, M. (2015). The ageing population and smart metering: A field study of householders' attitudes and behaviours towards energy use in Scotland. *Energy Research & Social Science, 9,* 107–115.

Bennett, P., & Mandell, R. (1969). Prepurchase information seeking behavior of new car purchasers: The learning hypothesis. *Journal of Marketing Research, 6*(4), 430–433.

De Meyer, A. (1993). Management of an international network of industrial R&D laboratories. *R&D Management. 23*(2), 109–120.

DECC. (2012). *Smart meters: Research into public attitudes*. London: Department of Energy & Climate change.

Farmer, J., & Foley, D. (2009). The economy needs agent-based modelling. *Nature, 406,* 685–686.

Galán, J., Izquierdo, L., Izquierdo, S., Santos, J., del Olmo, R., López-Paredes, A., & Edmonds, B. (2009). Errors and artefacts in agent-based modelling. *Journal of Artificial Societies and Social Simulation, 12.* http://jasss.soc.surrey.ac.uk/12/1/1.html.

Gereffi, G., Humphrey, J., & Sturgeon, T. (2005). The governance of global value chains. *Review of International Political Economy, 12*(1), 78–104.

Haney, A., Jamasb, T., & Pollitt, M. (2009). *Smart metering and electricity demand: Technology, economics and international experience* (EPRG Working Paper). Cambridge: University of Cambridge.

Hansen, M., & Birkinshaw, J. (2007). The innovation value chain. *Harvard Business Review, 86*(6), 121–130.

House of Commons. (2009). Engineering: Turning ideas into reality. Innovation: Universities.

Jager, W. (2000). *Modelling consumer behaviour*. Doctoral thesis, University of Groningen, Centre for Environmental and Traffic psychology, Groningen, Netherlands. http://docserver.ub.rug.nl/eldoc/dis/ppsw/w.jager/.

Janssen, M. A., & Jager, W. (1999). An integrated approach to simulating behavioural processes: A case study of the lock-in of consumption patterns. *Journal of Artificial Societies and Social Simulation*, *2*(2).
Karahanna, E., Straub, D., & Chervany, N. (1999). Information technology adoption across time: A cross-sectional comparison of pre-adoption and post-adoption beliefs. *MIS Quarterly*, *23*(2), 183–213.
Keirstead, J., & Schulez, N. (2010). London and beyond: Taking a closer look at urban energy policy. *Energy Policy*, *38*, 4870–4879.
MacBryde, J., Paton, S., & Clegg, B. (2013). Understanding high-value manufacturing in Scottish SMEs. *International Journal of Operations & Production Management*, *33*(11/12), 1579–1598.
Martiskainen, M., & Ellis, J. (2009). *The role of smart meters in encouraging behavioural change—prospects for the UK* (Working Paper). Unset.
NAE. (2008). *The offshoring of engineering: Facts, unknowns, and potential implications*. Washington, DC: National Academy of Engineering National Academies Press.
Normann, R., & Ramirez, R. (1993). From value chain to value constellation: Designing interactive strategy. *Harvard Business Review*, 65–77.
Office for National Statistics. (2011). *Neighbourhood Statistics*. Available at: http://www.neighbourhood.statistics.gov.uk/dissemination/LeadKeyFigures.do;jsessionid=X1kmTPmcJGFfS8RG1sMMphhQDL2bj1YlJv1sqQh5qvG5FhvGsdRX!1475111031!1322231356285?a=3&b=276810&c=leeds&d=13&e=13&g=382985&i=1001x1003x1004&m=0&r=1&s=1322231356285&enc=1&nsjs=true&nsck=true&nssvg=false&nswid=1920.
Owens, S., & Driffill, L. (2008). How to change attitudes and behaviours in the context of energy. *Energy Policy*, *36*, 4412–4418.
Parthasarathy, M., & Bhattacherjee, A. (1998). Understanding post-adoption behavior in the context of online services. *Information Systems Research*, *9*(4), 362–379.
Porter, M. (1985). *Competitive advantage: Creating and sustaining superior performance*. New York: The Free Press.
Rixen, M., & Weigand, J. (2014). Agent-based simulation of policy induced diffusion of smart meters. *Technological Forecasting and Social Change*, *85*, 153–167.
Russell, S., & Norvig, P. (2003). *Artificial intelligence: A modern approach* (2nd ed.). Upper Saddle River, NJ: Prentice Hall.
Schiffman, L., Lazar, L., & Hansen, H. (2008). *Consumer behaviour, a European outlook*. Harlow: Pearson Education Limited.

Science and Skills Committee, Fourth Report of Session 2008–09. London: House of Commons.

Sharma, K., & Saini, L. M. (2015). Performance analysis of smart metering for smart grid: An overview. *Renewable and Sustainable Energy Reviews, 49,* 720–735.

Siebers, P., & Aickelin, U. (2011). A first approach on modelling staff proactiveness in retail simulation models. *Journal of Artificial Societies and Social Simulation, 14*(2), 2. Available at: http://jasss.soc.surrey.ac.uk/14/2/2.html.

Spiller, J., Vlasic, A., & Yetton, P. (2007). Post-adoption behavior of users of Internet service providers. *Information & Management, 44,* 513–523.

Thamhain, H. (1992). *Engineering management: Managing effectively in technology-based organisations.* New York: Wiley.

Tripathy, A., & Eppinger, S. (2011). Organizing global product development for complex engineered systems. *IEEE Transactions on Engineering Management, 58*(3), 510–529.

Xu, P., Shen, J., Zhang, X., Zhao, X., & Qian, Y (2015). Case study of smart meter and in-home display for residential behaviour change in Shanghai, China. *Energy Procedia, 75,* 2694–2699.

Zhang, Y., & Gregory, M. (2011). Managing global network operations along the engineering value chain. *International Journal of Operations & Production Management, 31*(7), 736–764.

Zhang, Y., Gregory, M., & Shi, Y. (2007). Global engineering networks (GEN): The integrating framework and key patterns. *Proceedings of the Institution of Mechanical Engineers, Part B: Journal of Engineering Manufacture, 221,* 1269–1283.

Zhang, Y., Gregory, M., & Shi, Y. (2008). Global engineering networks (GEN): Drivers, evolution, configuration, performance, and key patterns. *Journal of Manufacturing Technology Management, 19*(3), 299–314.

Zhang, Y., Gregory, M., & Shi, Y. (2014). Managing global engineering networks (GEN) PART I: Theoretical foundations and the unique nature of engineering. *Proceedings of the Institution of Mechanical Engineers, Part B: Journal of Engineering Manufacture, 228*(2), 163–171.

Zhang, T., Siebers, P., & Aickelin, U. (2011). Modelling electricity consumption in office buildings: An agent based approach. *Energy and Buildings, 43* (10), 2882–2892.

Zhang, T., Siebers, P., & Aickelin, U. (2012). A three-dimensional model of residential energy consumer archetypes for local energy policy design in the UK. *Energy Policy, 47,* 102–110.

Authors' Biography

Tao Zhang is Lecturer of Marketing and Sustainability at the Business School of the University of Birmingham, UK. He gained his PhD from the Energy Policy Research Group, Judge Business School, University of Cambridge, and has over ten years' of experience in innovation and energy research. His research outputs appear in marketing, energy, innovation and technology management journals.

Yufeng Zhang is Senior Lecturer of Operations Management at the Business School of the University of Birmingham, UK. He had an earlier industrial career with responsibities for project management and new business development, and continuees to work very closely with industry aiming to help companies cope with challenges and opportiunities in managing complex global engineering networks. He has researched and published in the area of engineering management, network operation, international manufacturing, technology management and innovation.

8

ICT for High Value Engineering Networks

Ting He and Xiaofei Xu

Eric Emerson Schmidt, the former Executive Chairman of Google once claimed that *"Technology is transforming virtually every business sector. This transformation is happening at an unprecedented pace, and it is accelerating"*. By reviewing the recent history of *industrial change*, we can see that science and technology have played a dominant role in the development of the modern economy and society. In particular, the profound transformations related to productivity, production methods and technology, business models and innovation models, initiated by the combination of the engineering industry and the information and telecommunication technologies (ICTs), are leading to the upgrading and transforming

T. He (✉)
College of Computer Science and Technology, Huaqiao University, Xiamen, China
e-mail: xuantinghe@hit.edu.cn

X. Xu
School of Computer Science and Technology, Harbin Institute of Technology, Harbin, China
e-mail: xiaofei@hit.edu.cn

Y. Zhang and M. Gregory (eds.), *Value Creation through Engineering Excellence*,
https://doi.org/10.1007/978-3-319-56336-7_8

towards high value engineering. Among various engineering fields, manufacturing is without doubt remarkable in the history of industrial development around the world.

1 Technology-Driven Industrial Evolution

1.1 The History of Industrial Evolution

Industrial revolution can also be seen the revolution of science and technology. Differing from the popular theory in America stating that there are three industrial revolutions, the progressiveness of the manufacturing technology can be described as a four-stage industrial revolution, involving the evolutionary processes from Industrie 1.0 to Industrie 4.0. The characteristics of the four industrial revolutions identified by Acatech (National Academy of Science and Engineering of Germany) are shown in Fig. 1 (Acatech 2013).

- *Industrie 1.0*: The first industrial revolution that mobilized the mechanization of production using water and steam power, starting from 1760s to the mid-nineteenth century. The achievements of this industrial revolution are mechanical production taking the place of

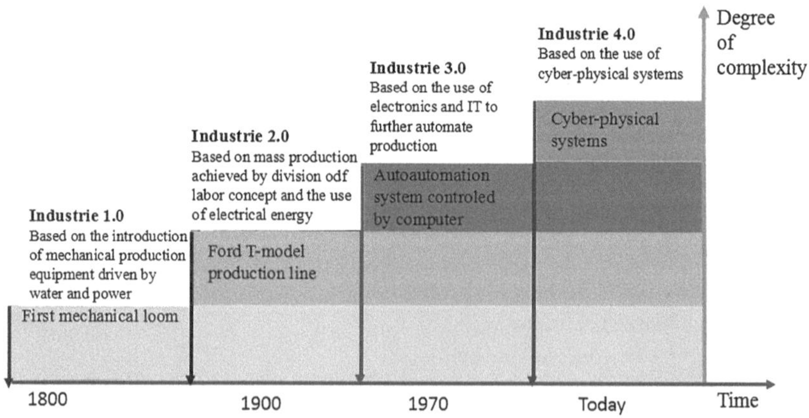

Fig. 1 The roadmap from Industrie 1.0 to Industrie 4.0

manual labour, and the economy transformed from that based on agriculture and handicraft industry to the model in which economic development was driven by industries and mechanical manufacture. Industrie 1.0 established the solid material foundation and the advanced technology base for human civilization.

- *Industrie 2.0*: The second industrial revolution introduced mass production with the help of electric power on the basis of division of labour starting from late nineteenth century to early twentieth century. Industrie 2.0 created a new pattern—sometimes called Fordism—which was standardized manufacturing and mass production by separating parts manufacturing and products assembling successfully.
- *Industrie 3.0*: The third industrial revolution started from 1970s and extended to about 2010. The wide application of ICTs contributed to the production revolution of automation and informatization of manufacturing process—sometimes called post-Fordism or flexible specialization. At this period, factories adopted mechanical equipment widely which are controlled automatically by real electronic and information technology such as PC, PLC/single-chip microcomputer, etc. This led to production efficiency, product quality, division of labour and mechanical equipment life had been improved unprecedentedly. From then on, machines gradually replaced people, this not only took over a great proportion of "physical labour", but also took charge of some "mental labour". It can be called Industrie 3.0 that makes manufacturing automated and informationalized continuously.
- *Industrie 4.0*: The fourth industrial revolution has been defined as involving technologies and concepts of value chain organization which draws together Cyber-Physical Systems, the Internet of Things and the Internet of Services.

German academics and industrialists believe that intellectualization based on Cyber-Physical System (CPS) will bring the fourth industrial revolution led by intelligent manufacturing in the next decade. The integration of product lifecycle, digitization of whole manufacturing process and ICT-based modules will establish a highly flexible, personalized and digitized production mode of product and service. The USA, China, Japan and other countries have also put forward their own understandings, definitions and development goals about Industry 4.0. This presents us with a new industrial blueprint: in an "intelligent, networked world", Internet of Things and Internet of Services will penetrate into all key social area. It will change the value creation process gradually, reorganize the industrial chain, and the traditional industry boundaries will disappear while a variety of new areas and cooperation forms will emerge.

1.2 ICT-Driven Industrial Evolution

If Industrie 1.0 and Industrie 2.0 are believed to adopt the steam engine and electric power as the driving force, respectively, the Industrie 3.0 and Industrie 4.0 are typical IT revolutions which employ the computer technology and communication technology as the driving force. These two are the most powerful industrial revolutions in the human history, which are rapidly transforming society from industrialization era, through automation, informatization era to the final service era.

To look back at history, since the first computer came into the world in 1946, almost every vital and significant innovation in ICT has been bringing new revolutions for the industrial society based on manufacturing (shown in Fig. 2).

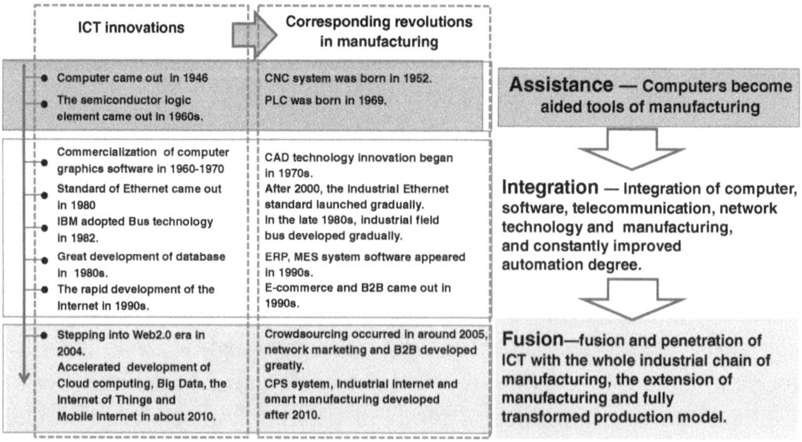

Fig. 2 ICT-driven development from Industrie 3.0 to Industrie 4.0

In the era of Industrie 3.0, ICT's development mainly experienced two stages. The first stage, involving the common application of mainframe computers and simple dumb terminals before 1980s, is regarded as the first generation of information technology platform. From the middle period of 1980s to the beginning of twenty-first century, the personal computers and distributed servers connected via the Internet were of great popularity and are considered to be the second generation of information technology platform. Based on the two stages of the technology platform, manufacturers had gradually realized the automation and digitization of production process and the informatization of management, leading to the emergence of the distributed, global division and supply chain system.

In the recent 5 years, the third generation of IT platform characterized by the mobile Internet, social networking, cloud computing, big data, Internet of Things has been developing dramatically. The emphasis of the third-generation IT platform is not the vertical upgrading of its various branches, but the horizontal penetration of information technology into the manufacturing, finance, culture and almost all industries. The "Internet +" with the main objective of deep integration of informatization and industrialization is the embodiment of the new generation of information technology (Chinese Central Government 2015) (Fig 3).

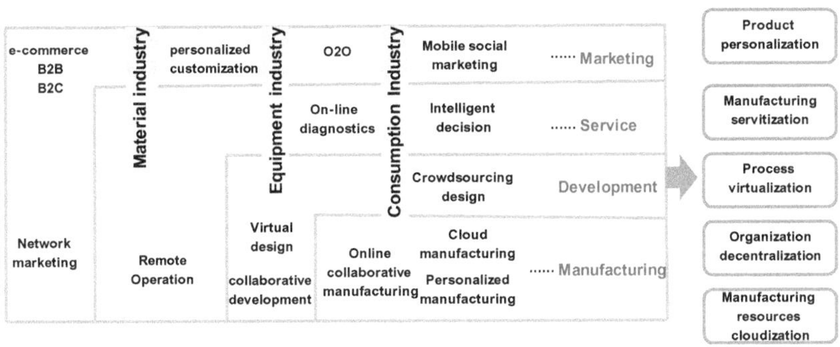

Fig. 3 The integration of the new-generation ICT and manufacturing

The integration of manufacturing and the Internet has been growing rapidly. The opening up of manufacturing enterprises has been increasing, and the integration degree of business processes has been increasingly deep. Starting from marketing, along the reverse supply chain business process flow, this revolution is transforming and upgrading the whole manufacturing industry and restructuring its value chain. This development makes manufacturing industry move towards the high value manufacturing/engineering distinguished by network, digitization, intellectualization, servitization, personalization and high flexibility. Further, it will also lead to revolutionary changes to the original value creation system and create a large number of innovative business models and promote the emergence and development of new industrial value ecosystem.

2 ICT-Enabled Advanced Manufacturing Technologies and Models

Advanced manufacturing is the use of advanced technology to improve products or processes. This sector mainly introduces the ICT-enabled typical advanced manufacturing technologies and models which are representative in manufacturing.

2.1 ICT-Enabled Advanced Manufacturing Technologies

Numerical control (NC) is the technology which allows machinery or other devices to work automatically by digital programming. In 1949, the first numerical control machine tool in the world was invented by the Parson Corporation and the Institute of Servo Mechanism in MIT jointly (Reintjes 1991). Its emergence promoted the development of machining automation technology, and it turned over the new page of Industrie 3.0 of manufacturing. In 1970s, the emergence of computer control technology accelerated the development of NC machine tool, and there appeared successively computer numerical control

(CNC) system (with the function of man–machine graphical interaction) and CAD/CAM technology. The development of the information super-highway and the Internet brought global change from 1990s to the first few years in early twenty-first century, and the progress of Industrial Ethernet, microelectronic technology, sensor technology, precision machinery technology, automatic control technology and personal computer technology, promoted the manufacturing technology represented by NC machine tool to a new higher level. In this process, a large variety of advanced manufacturing technologies such as CAX/DFX, computer integrated manufacturing (CIM), virtual manufacturing/simulation, etc., emerged.

In recent 5 years, the rocketing development of emerging computer technologies such as clouding computing, Internet of Things, big data and mobile Internet, etc., has triggered a new round of manufacturing technology innovation. 3D printing, industrial robotics, virtual reality, Cyber-Physical System, etc., are the representatives of the newest development achievements of advanced manufacturing technology at this time.

1. *3D printing*, (Wikipedia, 2016a) also known as additive manufacturing, refers to various processes used to synthesize three-dimensional objects. In 3D printing, successive layers of material are formed under computer control to create an object. These objects can be of almost any shape or geometry and are produced from a 3D model or other electronic data sources. A 3D printer is a type of industrial robot.

   3D printing in the term's original sense also refers to processes that sequentially deposit material onto a powder bed with inkjet printer heads. More recently, the meaning of the term has expanded to encompass a wider variety of techniques such as extrusion and sintering-based processes. Technical standards generally use the term additive manufacturing for this broader sense.
2. *Industrial robotics* An industrial robot is defined by ISO 8373 as an automatically controlled, reprogrammable, multi-purpose manipulator programmable in three or more axes. The field of robotics may be more practically defined as the study, design and use of robot systems for manufacturing. Typical applications of robots include welding,

painting, assembly, pick and place (such as packaging, palletizing and SMT), product inspection, and testing; all accomplished with high endurance, speed and precision.

3. *Virtual reality (VR)* (Wikipedia, 2016b) VR can be referred to as immersive multimedia or computer-simulated reality, which replicates an environment that simulates a physical presence in places in the real world or an imagined world, allowing users to interact in that world. VR artificially creates sensory experiences, which can include sight, touch, hearing and smell.

Most up-to-date virtual realities are displayed either on a computer screen or with special stereoscopic displays, and some simulations include additional sensory information and focus on real sound through speakers or headphones targeted towards VR users. Some advanced haptic systems now include tactile information, generally known as force feedback in medical, gaming and military applications. Furthermore, VR covers remote communication environments which provide virtual presence of users with the concepts of telepresence and telexistence or a virtual artefact (VA) either through the use of standard input devices such as a keyboard and mouse, or through multi-modal devices such as a wired glove or omnidirectional treadmills. The simulated environment can be similar to the real world in order to create a lifelike experience (for example, in simulations for pilot or combat training), or it can differ significantly from reality (such as in VR games).

4. *Cyber-Physical Systems (CPS)* CPS is a multi-dimensional intelligent technology system, which integrates computing, network and physical environment. Through core technical means (such as intelligent sensing, analysis, mining, evaluation, prediction, optimization, coordination, etc.) and the organic integration and depth cooperation of 3C (computing, communication and control) technologies, CPS can realize real-time sensing, dynamic control and information service of large-scale engineering system. The integrated design of the computing, communication and physical system of CPS can make engineering system more reliable, efficient and real-time collaborative.

In all the strategies of Industry 4.0 of Germany, Industrial Network of the U.S. (Peter and Macro 2012), and the "Made in China 2025"

(Chinese Academy of Engineering 2015), CPS is their common point and core element. CPS is the key to research and implementation of Industry 4.0 and other manufacturing strategies.

2.2 ICT-Based Advanced Manufacturing Models

Advanced manufacturing models are schemes for production and production organization that manufacturing industries use it to improve product quality, market competitiveness, scale of production and production efficiency. Since 1980s, with the extensive application of advanced manufacturing technology, more and more advanced manufacturing models and methods have been presented and found their large-scale application by academia and industry. Typical ones include CIM, concurrent engineering (CE), product data management/product lifecycle management (PDM/PLM), enterprise resource planning/supply chain management (ERP/SCM), agile manufacturing (AM), networked manufacturing (NM), manufacturing grid (MGrid), application service provider (ASP), virtual manufacturing (VM), service-oriented manufacturing (SOM), cloud manufacturing (CMfg), product service system (PSS), intelligent logistics, and smart manufacturing (SM).

1. Computer Integrated Manufacturing (CIM)

CIM was developed by Joseph Harrington in the early 1970s and released by his book "Computer Integrated Manufacturing". CIM is the integration of total manufacturing enterprise by using computer and communication systems coupled with new managerial philosophies that improve organizational and personnel efficiency. This term possesses the following features:

(a) CIM is a kind of philosophy of organization, management and production operation of enterprise. Its purpose is to make enterprises have high-quality product, short time-to-market, low cost, good service and success in competition. CIMS (computer integrated manufacturing system) is the realization of the philosophy.

(b) All the processes of enterprise production and operation management (including market analysis, business decision-making, management, product design, process planning, manufacturing, sales, after-sales and other business processes), are an integrated entity which should be coordinated from the point of view of the system as a whole to achieve the optimization globally.
(c) Enterprise's production elements include people, technology and management. In particular, people play the leading role in the production of modern enterprises.
(d) Business processes of the enterprise include information flow (collection, transmission and processing of information) and logistics. Modern enterprises should pay particular attention to the management of information flow and the integration of information flow and logistics.
(e) CIM is a comprehensive set of technologies which synthesizes and develops the computer-aided technologies in the production processes of enterprise, including computer-aided management technology (ERP/SCM), computer engineering design and management technology (CAD/CAM/CAPP/PDM), computer-aided quality management and control technology (CAQ) and computer-aided manufacturing technology, etc.

2. Agile manufacturing (AM)

According to Gupta and Mittal (1996), AM is "a business concept that integrates organization, people and technology into a meaningful unit by deploying advanced information technologies and flexible and nimble organizational structures to support highly skilled, knowledgeable and motivated people". In fact, companies which utilize an AM approach tend to have very strong networks with suppliers and related companies, along with numerous cooperative teams which work within the company to deliver products effectively. They can retool facilities quickly, negotiate new agreements with suppliers and other partners in response to changing market forces, and take other steps to meet customer demands. This means that the company can increase output of products with a high consumer

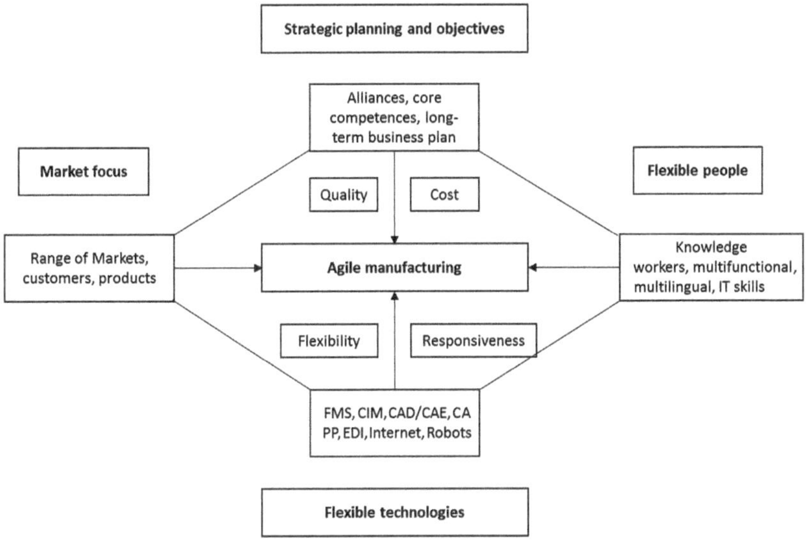

Fig. 4 The agile manufacturing paradigm. (Adapted from Luo et al. 1994)

demand, as well as redesign products to respond to issues which have emerged on the open market.

Studies have identified four principles of agility in AM: organizing to master change, leveraging impact of people and information, cooperating to enhance competitiveness and enriching the customer. Figure 4 represents the model of explaining the AM paradigm.

3. Service-oriented Manufacturing (SOM)

All manufacturers offer services, but some use services as the basis of their competitive strategy and this leads to the servitization of manufacturing (Baines et al. 2010). Servitization is now widely recognized as an innovation of a manufacturer's capabilities and processes. It is used to better create mutual value, through a shift from selling product to selling product–service systems. This novel manufacturing model—SOM—is viewed as a new industry paradigm of the integration of the production-based product economy and the consumption-based service economy.

The emergence of SOM is to realize the added value for stakeholders in manufacturing value chains (Lin et al. 2011). Through the integration of products and services, it can realize the integration of distributed manufacturing resources and high-degree synergism of their individual core competitiveness to achieve an innovative and efficient manufacturing mode.

(a) From the perspective of value, SOM emphasizes the transformation from traditional product-centric manufacturing to one providing products with rich services or product-based services.
(b) From the perspective of the production model, SOM emphasizes the transformation from a product-centric to a people-centric approach, and the integration of customer, operator's cognition and their knowledge. It can realize personalized production and service through mining the demand on the service manufacturing chain effectively;
(c) As regard to the organizational model, though SOM transcends the conventional manufacturing and service division, it pursues that different types of subjects (customer, service company and manufacturer) actively participate in the collaborative service-oriented manufacturing network. In the dynamic mutual collaboration, optimal allocation of resources can be made and then SOM system with dynamic stable structure will emerge.
(d) As regard to the operations model, SOM emphasizes active service and takes the initiative to allow customers to participate in product manufacturing and application services process, and discovers customer needs to provide targeted services. Based on business process cooperation between enterprises, it can provide productive service and service production for the customers from the upstream to the downstream of a supply chain and achieves collaborative value creation (Fig 5).

4. Cloud manufacturing (CMfg)

Cloud manufacturing is a service-oriented, knowledge-based smart manufacturing system with high efficiency and low energy consumption. As an emerging manufacturing model, it is developed from existing

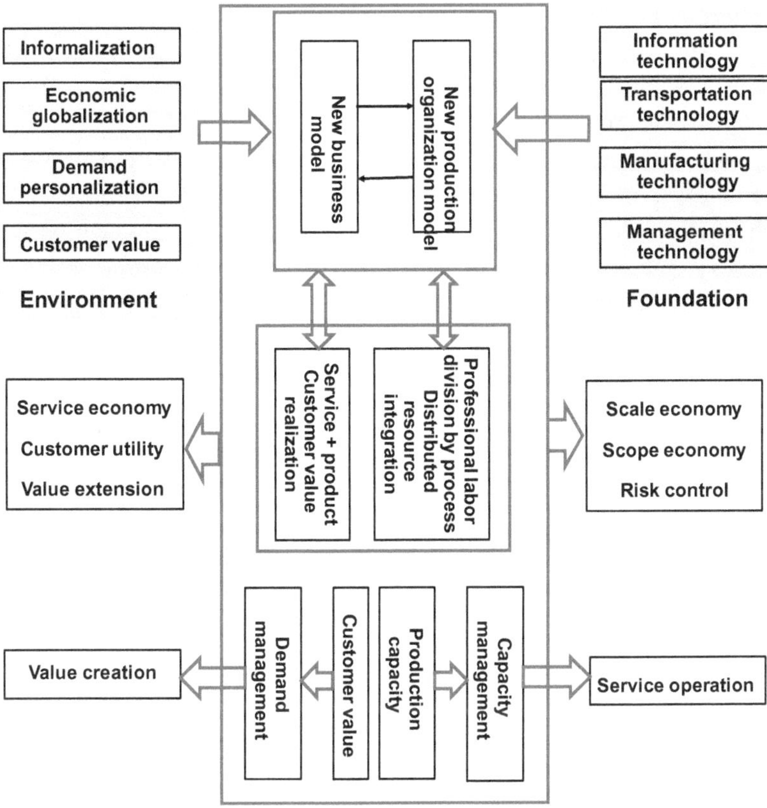

Fig. 5 SOM system structure. (Adapted from Li et al. 2010a, b)

advanced manufacturing models (e.g. ASP, NM, MGrid) and enterprise information technologies under the support of cloud computing, Internet of Things (IoT), virtualization and service-oriented technologies, etc. It extends and transforms manufacturing resources and manufacturing capabilities into manufacturing services, which can be managed and operated in an intelligent and unified way to enable the full sharing and circulating of manufacturing resources and manufacturing capabilities. CMfg can provide safe and reliable, high-quality, cheap and on-demand manufacturing services for the whole lifecycle of manufacturing which includes pre-manufacturing (argumentation, design, production and sale), manufacturing (product usage, management and maintenance), and post-manufacturing (dismantling, scrap, and recycling). The concept of cloud manufacturing was initially proposed by Professor Bohu Li in (2010a, b) (Fig 6).

5. Smart Manufacturing (SM)

"Smart manufacturing" was firstly presented by Wright P.K. and Bourne D.A. in their book of "Manufacturing Intelligence" in 1988. With the development of ICT and manufacturing engineering, SM becomes a generic term about advanced manufacturing process, system and model, which is based on the new ITs. It includes all the processes of design, production, management, service and other manufacturing activities and has the functions of deep information self-perception,

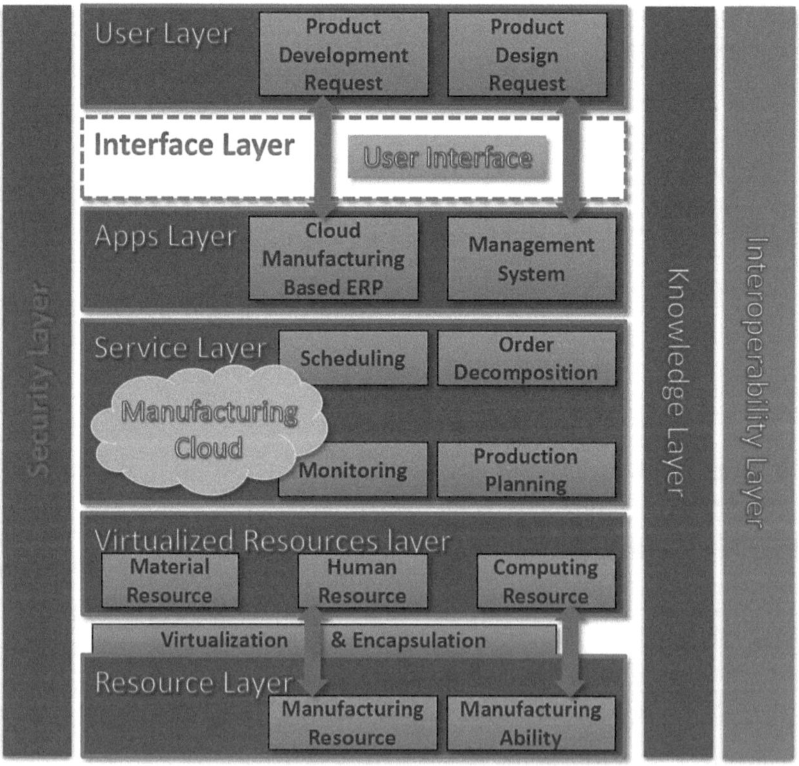

Fig. 6 Architecture of cloud manufacturing system. (Adapted from Li et al. 2010a, b)

intelligent optimized self-decision, precise controllable self-execution, etc. SM updates the concept of manufacturing automation and extends it to the flexible, intelligent and highly integrated one, and will continue to extend and expand it over time.

The essence of SM is the mutual penetration of virtual network and physical production. On the one hand, information network will completely change the model of production organization to realize large-scale personalized service manufacturing. On the other hand, manufacturing will be an extension and node of the Internet to expand the scope and effect of the network economy. Smart manufacturing, the latest forms of manufacturing supported by the network interconnection and taking the smart factory as its carrier, can effectively shorten the lead time of product development, improve production efficiency, promote product service value, reduce resource and energy consumption. From the combination of software and hardware perspectives, smart manufacturing is a manufacturing system of “virtual network plus physical entity”. The strategies of the Internet of Industrial Things, the industry 4.0, and the “Internet plus” described below later reflect the characteristics of smart manufacturing which are the deep integration of virtual network and physical entity.

3 Trends of the HVE in Major Economies of the World

Since the beginning of the twenty-first century, manufacturing has been confronted with opportunities and challenges brought by global industrial restructuring. Especially after the international financial crisis in 2008, various nations realized that only the real economy was the key to sustainable development. In order to stimulate their economic growth and reshape their competitiveness, many developed countries have implemented a series of national strategies related to manufacturing. These strategies include: the “Revitalize American Manufacturing Act” (2009), “Advanced Manufacturing Partnership” (2011), the “National Strategic Plan for Advanced Manufacturing” (2012), “National Network for Manufacturing Innovation: A Preliminary Design” (2013) of America, “Industry 4.0” (2013) of Germany, “High Value Manufacturing” and

"UK Industry 2050" of the UK (2013), "Europe 2020: A strategy for Smart, Sustainable and Inclusive Growth" (2010) of the European Commission, "Made in China 2025" (2015), "日本再興戦略" (2013 and 2014), "New growth power industry planning and development strategy" (2009) and "Future growth momentum to implement the plan" (2014) of Korea, "Make in India" of India (2014), etc.

3.1 Germany

Industrie 4.0 is a "strategic initiative" of the Germany government that was adopted as part of the High-Tech Strategy 2020 Action Plan, which was presented by the National Academy of Science and Technology of Germany on 8 April 2013 at the Hannover Fair. This initiative is intended to take up a pioneering role in industrial IT, which is currently revolutionizing the manufacturing engineering sector supported by the academia and industry of Germany, and it is intended to allow Germany to stay a globally competitive high-wage economy.

Industrie 4.0 is the German vision for the fourth industry revolution, where smart factories use ICTs and Cyber-Physical System to digitize and intellectualize their processes and reap huge benefits in the form of improved quality, lower costs and increased efficiency. In order to bring about the shift from current industrial production to Industrie 4.0, it is stated that Germany needs to adopt a dual strategy: (1) Germany's manufacturing equipment industry should seek to maintain its global market leadership by integrating ICT into its traditional high-tech strategies so that it can become the leading supplier of smart manufacturing technologies; (2) At the same time, it will be necessary to create and serve new leading markets for CPS technologies and products. In order to deliver the goals of this dual CPS strategy, the following features of Industrie 4.0 are envisaged:

- Horizontal integration through value networks
- End-to-end digital integration of engineering across the entire value chain
- Vertical integration and networked manufacturing systems

Under the new generation of ICT, the mission of Industrie 4.0 is to establish a highly flexible and digital production paradigm of product and service. The traditional industry boundaries will disappear, and accordingly various brand new business models will emerge.

The system of Industrie 4.0 presented by Germany comprises new technologies such as the Internet of Things (IoT), CPS, Internet of Services (IoS), etc., and it expects that the novel value creation system of manufacturing (driven by the IoT and the IoS) will be formed gradually and include the entire process from material (supply chain), plant to delivery. The following aspects characterize the vision for Industrie 4.0 (Fig 7):

As a key component of this vision, smart factories will be embedded in inter-company value networks and will be characterized by end-to-end engineering that encompasses both the manufacturing process and the manufactured product, achieving seamless convergence of the digital and physical worlds.

In the future under Industrie 4.0, it will be possible to incorporate individual customer- and product-specific features into the design, configuration, ordering, planning, production, operation and recycling phases. It will even be possible to incorporate last-minute requests for changes immediately before or even during manufacturing or operation.

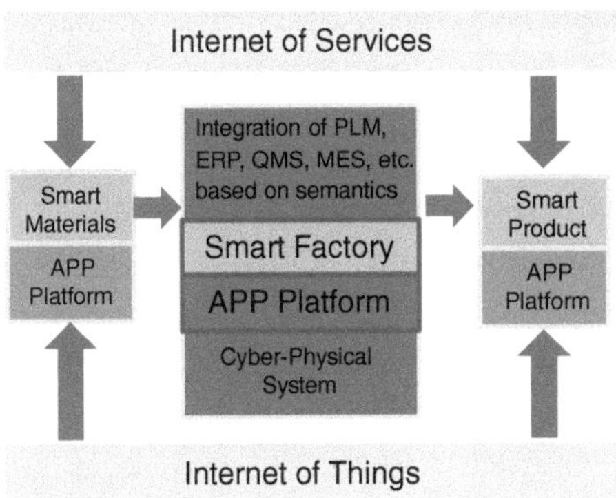

Fig. 7 The framework of Industrie 4.0

Implementation of the Industrie 4.0 vision will enable employees to control, regulate and configure smart manufacturing resource networks and manufacturing steps based on situation- and context-sensitive targets. Employees will be freed up from having to perform routine tasks, enabling them to focus on creative, value-added activities. Industrie 4.0 will lead to the development of new business and partnership models that are far more geared towards meeting individual, last-minute customer requirements.

3.2 The USA

In 2006, the concept of Cyber-Physical System was presented by American policy-makers and viewed as the breakthrough point of the new-generation technology revolution. After the financial crisis of 2008, aiming to create jobs and encourage reshoring of manufacturing, the USA administration has once again started to attach greater priority to the advanced manufacturing sector and published a number of policies to re-plan its future manufacturing.

For example, President Obama signed "Manufacturing Promotion Act" in 2010 to reconstruct American manufacturing industry through the adoption of future science and technology such as digital manufacturing and artificial intelligence, etc. In the summer of 2011, the President also launched the Advanced Manufacturing Partnership (AMP), a private-sector-led body that brings together representatives of the research, business and political communities to chart a "course for investing and furthering the development of the emerging technologies". In Feb. 2012, the National Science and Technology Council (NSTC) released American national strategy of the "National Strategic Plan for Advanced Manufacturing". This plan introduced the challenges of American manufacturing and the trends of global advanced manufacturing, and put forward the five goals for implementing the strategy plan and the relevant countermeasures. In May 2014, the White House announced a plan to build the third innovative research union of the "Digital Manufacturing and Digital Innovation Institute" after the National Additive Manufacturing Innovation Institute. Around the CPS application, this institute focuses on improving the capability of digital design and manufacturing in the four core technologies of advanced manufacturing, smart machine, advanced analytics and network entity safety.

In order to adapt to these strategies, the company GE published "Industrial Internet: Pushing the Boundaries of Minds and Machines" in November 2012. This report presents the new concept of the "Industrial Internet" which in essence is a global open network connecting and combining the three elements of intelligent machines, advanced analytics and people at work. It describes the future roadmap for transforming and servitizing manufacturing and treats smart devices, intelligent systems, intelligent decision-making as the key elements of the Industrial Internet. To maximize the value of the new ecosystem, the Industrial Internet Consortium was founded in March 2014 to bring together the organizations and technologies necessary to accelerate the growth of the Industrial Internet by identifying, assembling and promoting best practice. Membership includes small and large technology innovators, vertical market leaders, researchers, universities and government organizations (Fig 8).

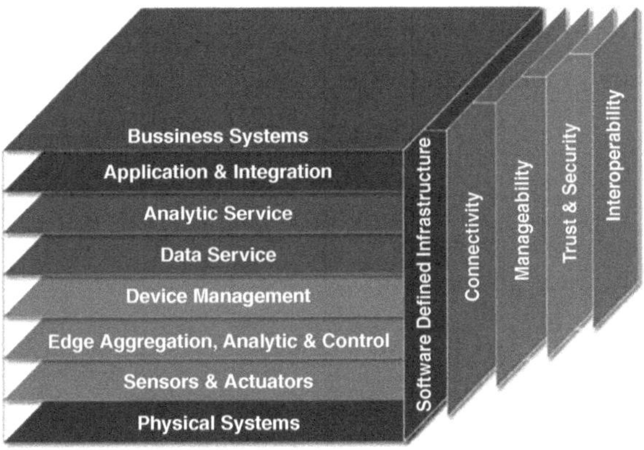

Fig. 8 Core technologies of software and Internet-based Industrial Internet Reference Architecture of IIC

3.3 China

At present, China has the most complete industrial system around the world. There exists a complete, independent industrial system which dominates Chinese GDP. The Chinese government attaches the great importance to the development and revitalization of manufacturing industry. Over the past several years, under the deep fusion of the new-generation ICT and manufacturing, the Chinese government has been committed to the restructure and upgrading of its economic growth model and industrial structure. It has launched a number of national strategic plans such as “the equipment manufacturing industry restructure and revitalization plan”, the “12th five-year plan for industrial transformation and upgrading”, “12th five-year development plan for intelligent equipment manufacturing industry” and the latest plans of “Internet Plus”, “Made in China 2025” (Chinese Industry 4.0).

The “Made in China 2025” should be a priority for all these above plans in the 13th Five-Year Plan. It is an exploration that conforms to Chinese actual industrial development and a development plan of manufacturing to promote automation, digitalization and intellectualization in the next 10 years. Its goals are to realize high-quality and high value manufacturing, green manufacturing, innovation-driven manufacturing and smart manufacturing in the ten key areas of new-generation IT (high-end numerical control machine and robot, aviation and aerospace equipment, oceaneering equipment and high-technology ship, advanced rail transportation equipment, energy-saving and new energy vehicles, electric power equipment, agricultural equipment, new materials, biological medicine and high-performance medical devices). In order to elaborate and refine the plan, the technology roadmap of the ten key areas for the plan was also proposed in November 2015.

In the “Made in China 2025”, intelligent manufacturing is positioned as the main target of Chinese manufacturing industry, and a reference structure model for Chinese intelligent manufacturing is developed from the perspectives of system architecture, value chain and product lifetime by China Electronics Standardization Institute (2015) (Fig 9).

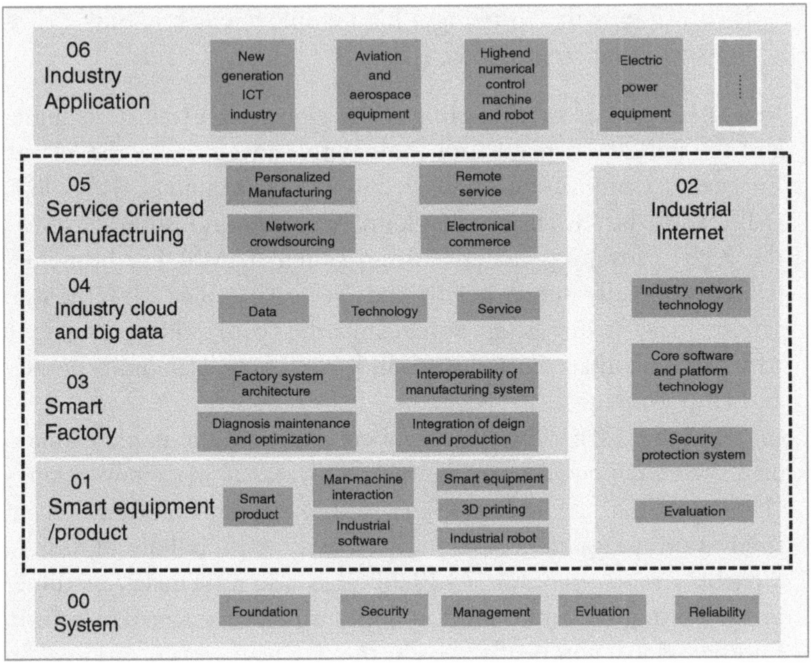

Fig. 9 The reference structure model for Chinese intelligent manufacturing

3.4 Eu

As the cradle of the modern industry, EU has played a leading role in global industrial development since the first industrial revolution. Since entering the second decade of the twenty-first century, it has successively promulgated several important development strategic plans below for developing advanced manufacturing industry.

1. In June 2010, the "EU 2020 Strategy" was launched to put forward three mutually reinforcing priorities:
    - Smart growth: Developing an economy based on knowledge and innovation.
    - Sustainable growth: Promoting a more resource efficient, greener and more competitive economy.

- Inclusive growth: Fostering a high-employment economy delivering social and territorial cohesion.

2. In addition, based on the status of the global high-tech development and the industry demand and comparative advantage in the EU, the European Commission identified six key enabling technologies (KETs) for the sustainable development of the industries. As one of the KETs, advanced manufacturing system (AMS) has been highlighted and fully supported by the EU with policy incentives and action plans, aiming at enhancing the competitiveness of the advanced manufacture industry and improving economic property and employment.
3. In December 2013, the EU members approved the implementation of a new framework programme for research and innovation of "Horizon 2020". Horizon 2020 is the biggest EU Research and Innovation programme ever, with nearly €80 billion of funding available over 7 years (2014–2020). It is also a financial instrument implementing the Innovation Union, an initiative aimed at securing Europe's global competitiveness in the five new areas of material and its efficient use, future food safety, advanced manufacturing and smart urban transportation.

3.5 UK

As the cradle of the first industrial revolution, manufacturing industry had brought the economic prosperity to UK for 300+ years. With the development globalization, UK has experienced loss of industrial employment since the 1980s. In the financial crisis of 2008, the UK's real economy has also been hit heavily, and this subsequently forced the UK government strive to rebalance the economy towards manufacturing industry.

In December 2011, the Advanced Manufacturing Supply Chain Initiative, worth up to £125 m, was established to improve the global competitiveness of UK advanced manufacturing supply chains. The initiative aims to help existing supply chains grow and achieve world-class standards, while encouraging major new suppliers to provide inward investment. The fund is intended to support projects in established UK advanced manufacturing sectors, such as aerospace, automotive and chemicals. It will also target newer growth areas where the UK is well placed to make an impact, such as renewables and other low-carbon sectors.

In October 2013, a project report "The future of manufacturing:a new era of opportunity and challenge for the UK" was published by Foresight in the Government Office for Science. This report stated that manufacturing was set to enter a dynamic new phase, driven by rapid changes in technology, new ways of doing business, and potential volatility around the price and availability of resources which would provide substantial opportunities for the UK. Looking ahead to 2050, this foresight project suggested that manufacturing was no longer just about "production"—making a product and then selling it. Manufacturers were increasingly using a wider "value chain" to generate new and additional revenue from pre- and post-production activities (service-oriented manufacturing), with technology playing a central role in driving change.

3.6 Japan

In June 1990, the Ministry of International Trade and Industry of Japan published a 10-year plan for intelligent manufacturing, and invested 150 billion yen on its research and experimentation. In 1994, Japan launched the project of International Collaboration in Advanced Manufacturing including global manufacturing, manufacturing knowledge system, distributed intelligent control system, etc. At the same time, eight major automobile manufacturers of Japan adopted intelligent manufacturing technologies, which emphasized the integration of automation, informatization and traditional manufacturing and integrating the automatic manufacturing system using computer technology.

In order to adapt to the needs of industry transformation and to maintain the leading position in robotics, the Japanese government issued the "Japan's Robot Strategy" in January 2015 which put forward three core goals: (a) becoming a global base for robot innovation; (b) the country with the most use of robots in the world; and (c) leading the world in the robots area. To maintain the leading position, this strategy argued that Japan needed to implement a revolutionary strategy for robots which involved the deep integration between robots and IT, big data, network and artificial intelligence, building the technology leadership for robot innovation, and creating a world-class robot application society to lead the development of robot in the IoT era.

In May 2015, the Ministry of Economy, Trade and Industry of Japan published the White Paper on Manufacturing Industries (Japanese Industry 4.0). This report identified the changes of manufacturing industry in an IoT society in Europe and the USA and proposed Japanese countermeasures as follows: on the one hand, to increase the investment of research and development in the fields of the new industrial revolution technology such as robotics, 3D printing, new material, new energy, etc.; on the other hand, to strengthen the management innovation and business model transformation to deal with the changes of new environment using the IoT, cloud computing, etc.

4 Summary

Karl Marx during his study on the first industrial revolution explicitly pointed out that the transformation of productivity alone doesn't represent revolution, instead an industrial revolution means the transformation of relations among productive forces. This proposition has been verified by the second and again the third industrial revolutions. It is highly possible that the predication will continue to be valid for the fourth industrial revolution in which the transformations of productivity and productive relations occur simultaneously. For example, a typical technical application was the mechanization in the Industrie 1.0 and that led to the emergence of workshop or plant and the working class. The Industrie 2.0 represented by electrification prompted the revolution of workshop transforming to flowshop and the emergence of the professional manager class. The Industrie 3.0 represented by informatization led to emergence of distributed, global factory and global division of labour. The ongoing industrial revolution may also conform to the rule in which large amounts of ICTs will be applied to the production processes and this will inevitably lead to smart manufacturing systems.

Industrie 4.0 will not only be a transformation of production methods, but also a profound transformation of social structures and economic growth models driven by the new-generation ICTs. In this ICT era, the input of production changes from raw materials such as petroleum and steel to big data; the workshop machines become smart equipment; and manufacturers can connect with consumers directly and personalize products in terms of their requirements. These have disrupted the existing business models and the traditional social structure. To summarize, Industrie 4.0 is restructuring business models as well as social structures which will inevitably lead to the transformation of how an individual behaves, how a business operates, and how an economy grows.

References

Acatech-National Academy of Science and Engineering of Germany. (2013). *Securing the future of German manufacturing industry: Recommendations for implementing the strategic initiative INDUSTRIE 4.0: Final report of the industrie 4.0 working group*.

Baines, T. S., Lightfoot, H., Benedettini, O., & Kay, J. M. (2010). The adoption of servitization strategies by UK-based manufacturers. *Proceedings of the Institution of Mechanical Engineers Part B: Journal of Engineering Manufacture, 224*(5), 815–829.

China Electronics Standardization Institute. (2015). *White paper on intelligent manufacturing standard system*.

Chinese Academy of Engineering. (2015). *Made in China 2025*.

Chinese Central Government. (2015). Available at: http://baike.baidu.com/link?url=6c9LCAzCKN3Hfi3dDG7P9wCEo5jUbJr_cKxMRwYKY_xyEG3VTjFm3_KakeXGpQ4uuijM3oXsk8hibwIrxI10W0KHfLBAtJ8xImzCaTdxFmEbPyCP9yt0wKGf0qPUeWZ7hEEAfk75UiiPnERSHshAR_3CVVvbpCVp8mhshXtzI7K. Accessed March 02, 2016.

Gupta, U. G., & Mittal, R. O. (1996). Quality, time, and innovation based performance measurement system for agile manufacturing. *Annual Meeting of the Decision Sciences Institute, 3*, 1511–1513.

Li, B. H., Zhang, L., Wang, S. L., Tao, F., Cao, J. W., Jiang, X. D., et al. (2010a). Cloud manufacturing: A service oriented networked manufacturing model. *Computer Integrated Manufacturing System, 16*(1), 1–7.
Li, G., Sun, L. Y., & Gao, J. (2010b). Architecture and implementation of service-embedded manufacturing. *Science & Technology Progress and Policy, 27*(7), 45–50.
Lin, J. L., Miao, R., Jiang, Z. B., & Wang, L. Y. (2011). Service-oriented manufacturing: The development trend of Chinese manufacturing industry. *Advanced Materials Research, 201–203*, 388–392.
Luo, Z. B., Wang, J. S., & Zhou, Z. Y. (1994). Agile manufacturing—21st century manufacturing enterprise strategy. *Chinese Journal of Mechanical Engineering, 30*(4), 1–5.
Peter, C., & Macro, A. (2012). *Industrial internet: Pushing the boundaries of minds and machines*. GE Report.
Reintjes, J. F. (1991). *Numerical control: Making a new technology*. Oxford: Oxford University Press.
Wikipedia. (2016a). *3D printing*. Available at: https://en.wikipedia.org/wiki/3D_printing.
Wikipedia. (2016b). *Virtual reality*. Available at: https://en.wikipedia.org/wiki/Virtual_reality.
Wright, P. K., & Bourne, D. A. (1988). *Manufacturing intelligence*. Reading, PA: Addison-Wesley.

Authors' Biography

Ting He is professor of Computer Science and Technology at Huaqiao University and previously at Harbin Institute of Technology, China. His research focusses on smart computing and services. He is currently exploring the data-enabled innovation of business model and business intelligence.

Xiaofei Xu is professor of Computer Science and Technology at Harbin Institute of Technology, China. He is President of Harbin Institute of Technology (Weihai). He chairs the Technical Committee on Service Computing of China Computer Federation.

9

Engineering Value Chain Modelling and Optimization

Lina Zhou and Xiaofei Xu

1 Background

The aim of this chapter is to introduce the modelling and optimization methods in engineering value chain decision-making and show the effectiveness and advances of solving management problems by information technologies. Decision-making is a very important part of engineering value chain management. There are many critical problems, such as selecting one among a number of suppliers, determining the order quantities of individual items in the next period, choosing the appropriate locations to set up the warehouses, allocating the product inventory among different distribution centres, deciding which route to take for the

L. Zhou (✉) · X. Xu
School of Computer Science and Technology,
Harbin Institute of Technology, Harbin, China
e-mail: zln@hit.edu.cn

X. Xu
e-mail: xiaofei@hit.edu.cn

Y. Zhang and M. Gregory (eds.), *Value Creation through Engineering Excellence*,
https://doi.org/10.1007/978-3-319-56336-7_9

transportation vehicle and so on and so forth. Based on the characters of a specific problem, different decision methods are developed and applied.

In this chapter, we summarize several typical value chain decision problems and related decision methods and introduce the popular decision-making methods based on mathematic model and optimization algorithms. We then investigate an engineering value chain construction decision problem, develop a multi-objective model, and propose the genetic algorithm-based solution procedure. Finally, numerical experiment and discussion are conducted to demonstrate the benefit of the method. Further trends of development in engineering value chain modelling and optimization will be illustrated.

2 Engineering Value Chain Decision-Making Problems and Methods

Problem solving is an essential skill for business and life. As an important problem-solving activity, the role of decision-making is to identify and choose alternatives based on the values and preferences of the decision-makers. The most popular decision-making problems in engineering value chain management can be categorized into four broad classes: partner selection, location-allocation, inventory decision, and vehicle routing.

1. Partner selection problem

The right partners must be selected to be involved in the value chain according to certain criteria. Partner selection is one of the key factors influencing the strategic alliance's performance, and the success of a value chain depends on the selection of the right partners (Beamish 1987). Developing appropriate criteria to evaluate the alternatives and select the right ones are the crux of partner selection. Many methods have been proposed and applied to different partner selection problem in the literature (Boer 2001; Chai 2013; Govindan 2015).

2. Location-allocation problem

Proposed by Cooper (1963), this location-allocation problem concerns the location and relation of a set of facilities that will provide homogeneous services. In engineering value chain management, the "facilities" could be plants, warehouses, distribution centres, service centres and any other nodes of an EVC. Location-allocation decision plays a very important role in the strategic design of value chain networks. Its decision object is to decide the best number and location of the facilities subject to certain constrains. Snyder (2006), Melo (2009), and Farahani (2012) reviewed the location-allocation decision problems from different perspectives.

3. Inventory decision problem

Inventory is the quantity of goods or materials in stock. As buffers to balance supply and demand, most of the nodes of the engineering value chain need to carry inventories. Inventory management is a very important part of core operations activities and will affect the performance of the value chain. The management of inventory requires a number of decisions. When to replenish stock, and how much to order are the two basic decisions in inventory management, which is the so-called inventory policy. Generally, the objective of inventory decisions is to decide the inventory policy that minimizes the total inventory cost, which include the ordering cost, carrying cost, and shortage cost. Uncertainty, revenue, and deterioration should also be considered in practical inventory decision problems (Bakker 2012; Horenbeek 2013).

4. Vehicle routing problem

One of the most important factors in implementing engineering value chain management is to efficiently control the physical flow of the material and goods. Vehicle routing decisions concern physical flows, which focus on how to effectively transport the materials and the products while minimizing the total cost. For example, how to transport the products to the final customers via the plants, warehouses, and distribution centres using group of vehicles. As a generalization of the travelling salesman problem, the vehicle routing formulation was first introduced by Dantzig and Ramser (1959). Beyond the classical formulation, a number of variants have been studied. Beyond this classical formulation, a number of variants have been proposed. Eksioglu (2009) presents a methodology for classifying the literature of the vehicle routing problem. Pillac (2013) gave a good survey of the dynamic vehicle routing problem. Lin (2014) reviewed the popular green vehicle routing problems.

The aforementioned problems are the most typical and popular decision problems in engineering value chain management, which have been hot topics in both academe and industry, and are receiving increasing attention. Based on the characters of specific decision problem, different kinds of decision methods are developed. Multi-criteria decision aid method, data mining technology-based method and optimization model-based method are the three commonly used methods in engineering value chain decision-making.

1. Multi-criteria decision-making methods

Multi-criteria decision-making methods are generally realized in the following paradigm: a decision-maker considers a set of alternatives and seeks to take an "optimal" decision considering all the factors that are relevant to the analysis. Developed to standardize the complex decision process and based on the alternatives evaluation theory, multi-criteria decision methods are very efficient in dealing with decision problems in which alternatives form a finite discrete set, typically consisting of a small number of elements, in which each alternative is fully known in complete detail, and any one of them can be selected as the decision. However, in many decision cases, the alternatives are complex, infinite, and not given in advance.

2. Data mining technology-based methods

Data mining is the computational process of discovering patterns in LARGE datasets involving methods at the intersection of artificial intelligence, machine learning, statistics, and database systems. The overall goal of the data mining process is to extract information from a data set and transform it into an understandable structure for further use. Discovering hidden knowledge from huge amounts of data will strongly improve the decision quality. For example, mining customer data can predict future demand and thus aid the production, inventory, and delivery decisions. Collecting large amounts of business data along an EVC, developing appropriate data cleaning and mining algorithms are key issues in applying data mining technology in decision-making.

3. Optimization model-based methods

Originated and defined in mathematics, an optimization problem concerns finding the best solution from all feasible solutions. It is applied to a widening array of contexts, including machine learning and information retrieval, engineering design, economics, finance, and management. Optimization model-based methods are very efficient in dealing with the complex constrained decision problems with infinite and unknown alternatives. Optimization model-based methods can identify the set of all possible alternatives and provide the optimal solutions for the decision-makers.

3 Decision-Making Based on Optimization Model and Algorithms

The standard form of an optimization model is formulated as follows (Boyd 2004):

$$\begin{aligned} & Minimize\, f(x) \\ & subject\ to \\ & \qquad g_i(x) \leq 0,\ i = 1, 2, \ldots, m \\ & \qquad h_i(x) = 0,\ i = 1, 2, \ldots, n \end{aligned} \tag{1}$$

where $f(x)$ is the objective function to be minimized over the variable x, $g_i(x) \leq 0$ are called inequality constraints, and $h_i(x) = 0$ are called equality constraints.

As shown in Fig. 1, the parameters represent the decision environment, the variables represent the candidate alternatives, the objective function(s) is (are) the criterion(s); the constraints are the requirements on the alternatives.

According to the number of the objectives, the optimization models can be divided into two categories: single-objective models and multi-objective models. The goal of single-objective models is to find out the best solution, which will lead to the minimum (maximum) value of the single-objective function. Cost is usually formulated as the single objective in many value chain management decision problems. Examples include Gumus (2009), Monteiro (2010), Başligil (2011), Creazza (2012), and Lee (2014). Depending on the form and functional description of the optimization problem, different optimization techniques can be used for the solution, linear programming, nonlinear programming, discrete optimization, etc. (Nemhauser 1989).

However, almost every real-world problem involves simultaneous optimization of several incommensurable and even competing objectives. In multi-objective optimizations, the various objective functions conflict with each other (i.e. optimizing one of them usually tends to move another towards undesirable values) and the aim is to simultaneously optimize a group of conflicting objectives. The interaction among

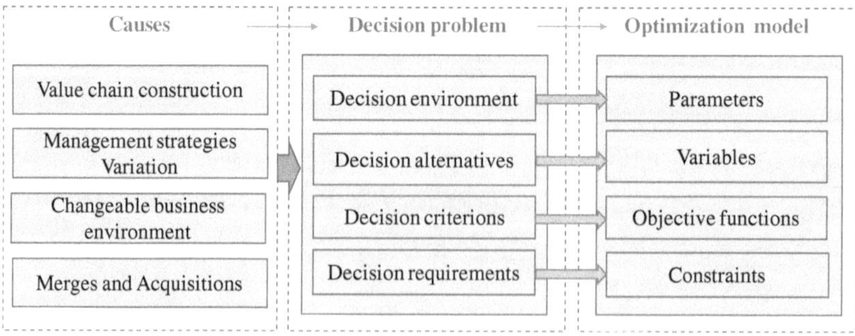

Fig. 1 Relationship between the decision problem and the optimization model

different objectives gives rise to a set of compromised solutions, largely known as the trade-off, non-dominated, non-inferior, or Pareto-optimal solutions. For multi-objective decision problems, there is no single optimal solution, but a set of alternative solutions. Pareto-optimality is expected to provide flexibility for the decision-makers.

Traditionally, there are several popular methods available in the operational research literature for solving multi-objective programming models, such as goal programming (Charnes 1957), goal attainment (Kiresuk 1968) and weighted sum method (Turban 1998). Due to the fact that multi-objective optimization problems are usually NP-hard, evolutionary algorithms are found efficient for solving multi-objective models (Zitzler 1999). Originated in the late 1950s, the term evolutionary algorithm stands for a class of stochastic optimization methods that simulate the process of natural evolution. Some famous evolutionary algorithms for multi-objective decision problems include vector evaluated genetic algorithm (VEGA) (Schaffer 1985), multi-objective genetic algorithm (MOGA) (Fonseca 1993), niched Pareto-genetic algorithm (NPGA) (Horn 1994), strength Pareto-evolutionary algorithm (SPEA) (Zitzler 1999), non-dominated sorting genetic algorithms (NSGA) (Srinivas 1994), Pareto-archived evolution strategy (PAES) (Knowles 1999), etc.

Efficient in handling alternative large search spaces and generating multiple alternative trade-offs, evolutionary algorithms are widely applied in solving multi-objective decision problems. Today, multi-objective decision is evolving towards the application of computer algorithms to solve mathematical models on computers.

4 Engineering Value Chain Configuration Model

In high value engineering industry (such as the aerospace industry), product design and fabrication is a long-term process and contains complex subprocesses that usually need the combined efforts of numerous organizations, ranging from very small enterprises to large corporations. Such value networks are usually very complex as they are composed of many nodes to handle many kinds of tasks. Thus, value network design and optimization is very complicated and hard to complete manually in high value engineering industry.

To interpret the decision process based on an optimization model and algorithm in detail, an engineering value chain construction problem is investigated, belonging to the location-allocation decision category and involves two companies—A and B. Suppose company A is an aerospace manufacturing company and has a mature supply network which is composed of the internal and external suppliers, warehouses, and assembly plants. Company B is a supplier of hardware and related components to aerospace original equipment manufacturers and their subcontractors and has its own distribution network composed of the plants and the regional warehouses, distributing the components to the customers.

The supply relationship before the acquisition is shown in Fig. 2. BP_{p} are the plants of company B, and BW_i represent its regional warehouses. S_{s} are the suppliers of company A (including both internal and external suppliers, indicated by different colours). AW_j represent company A's regional warehouses. C_k are company B's general customers, while AP_{q} are the assembly plants of company A, which are also customers of company B. M types of products are produced in company B's plants

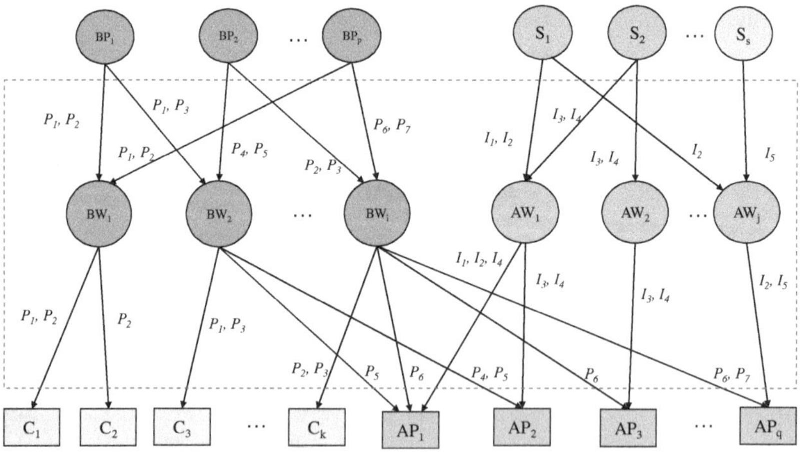

Fig. 2 Two value chain networks prior to acquisition

(BP_p) and distributed to the regional warehouses BW_i. Then, BW_i serve the customers (C_k and AP_q). Company A's suppliers (S_s) supply n types of items to company A's regional warehouses (AW_j), which distribute them to the right assembly plants (AP_q). The assembly plants integrate the parts and components supplied by company B and the suppliers into modules, subassemblies, and finally, aircrafts.

Before the acquisition, company B has to rent the warehouses (BW_i) and maintain its own distribution channel, which is a massive and costly process in order to satisfy the customers' (C_k and AP_q) requirements. It is obvious that company A's mature supply network can be utilized to distribute company B's products to the customers' (C_k and AP_q) after the acquisition. Thus, how to integrate the two multi-item value chain networks is a critical issue. The required network integration decisions include: (1) assessing whether to retain or eliminate company B's specific warehouses (BW_i); (2) warehouse expansion strategy and the new capacities of each warehouse (BW_i and AW_j); (3) supply links between the facilities; (4) the quantity of items shipped among the facilities.

A mathematical model will be developed to solve the network integration problem. The proposed model is based on the following assumptions:

1. Company A's suppliers (including internal and external suppliers) and company B's plants are retained, and their production capacities remain unchanged after the acquisition.
2. Company B's general customers are retained and their future product requirements are estimated based on the history business data.
3. All the requirement of company A's assembly factories must be satisfied.
4. Each regional warehouse of company B can be eliminated with a certain penalty cost.
5. Each regional warehouse of company B and A can be expanded with a certain cost.
6. Some important items must be stored in the appointed regional warehouses.

Consider that the new value chain network will distribute different kinds of items to different customers (general customers and company A's assembly plants) and the decision-makers generally compromise incompatible objectives to achieve adequate economic profits. We formulate the network integration problem as a multi-objective programming model with three objectives: (1) minimization of total integration cost, which consists of the costs of closing former warehouses, and adjusting the capacities of existing warehouses. (2) Minimization of total operation cost, including the annual fixed cost of warehouse operation, as well as the variable costs such as transportation cost. (3) Maximization of the general customer satisfaction rate, which is calculated by the weighted sum of each general customer's customer satisfaction rate.

Sets and indices

- I Index set of items; $i \in I$
- S Index set of the first-layer facilities (including firm B's plants and firm A's suppliers); $s \in S$
- W Index set of the second-layer facilities (including firm B's existing warehouse and firm A's existing warehouse); $j \in W$
- C Index set of the general customers; $k \in C$
- P Index set of firm A's assembly plants; $p \in P$
- M Index set for capacity levels available to existing warehouses; $m \in M$, in which $m = 0$ means that the warehouse is closed

Model parameters

- SC_{isj} Unit transportation cost of item i from S_s to $W_j s$
- TC_{ijn} Unit transportation cost of item i from W_j to C_n or P_n; $n \in C \cup P$
- FC_{jm} Fixed cost of operating W_j with capacity level m
- AC_{jlm} Cost of adjusting W_j capacity level from l to m; $l, m \in M$, in which $AC_{jl0} (l \neq 0)$ is the penalty cost of closing W_j with capacity level l, while $l = m$, $AC_{jlm} = 0$
- CR_{ik} Demand for item i of C_k
- PR_{ip} Demand for item i of AP_p

C_{is} Production capacity of item i for S_s; in which $C_{is} = 0$, while S_s does not produce item i

G_{jm} Capacity with level m for existing W_j

λ_n The relative importance of general customer; $n \in C$ and $\sum \lambda_n = 1$

α_{ij} The appointed storage relationship for item i; $\alpha_{ij} = 0$ or 1

Model variables

$$X_{jlm} = \begin{cases} 1 & \text{if the capacity level of } W_j \text{ is adjusted form } l \text{ to } m \\ 0 & \text{otherwise} \end{cases}$$

Particularly, $W_{jll}= 1(l \neq 0)$ indicates the unchanged W_j capacity level, and $W_{jl0}= 1(l \neq 0)$ indicates closure of an existing W_j with capacity level l.

Y_{isj} Amount of item i shipped from S_s to W_j

Z_{ijn} Amount of item i shipped from W_j to C_n or P_n; $n \in C \cup P$

With the above notation, the value chain network integration problem is formulated as follows:

$$\text{Min Cost}_I = \sum_{j \in W} \sum_{l \in M} \sum_{m \in M} AC_{jlm} X_{jlm} \tag{2}$$

$$\text{Min Cost}_O = \sum_{j \in W} \sum_{m \in M} FC_{jm} \sum_{l \in M} X_{jlm} + \sum_{i \in I} \sum_{s \in S} \sum_{j \in W} Y_{isj} SC_{isj} + \sum_{i \in I} \sum_{j \in W} \sum_{n \in C \cup P} Z_{ijn} TC_{ijn} \tag{3}$$

$$\text{Max PDC} = \sum_{n \in C} \lambda_n \left(\sum_{i \in I} \left(\left(\sum_{j \in W} Z_{ijn} \right) \Big/ CR_{in} \right) \right) \tag{4}$$

Subject to:

$$\sum_{l,m \in M} X_{jlm} \leq 1, \forall j \in W \tag{5}$$

$$\sum_{s \in S} Y_{isj} = \sum_{n \in C \cup P} Z_{ijn}, \forall i \in I, j \in W \tag{6}$$

$$\sum_{j \in W} Y_{isj} \leq C_{is}, \forall i \in I, s \in S \tag{7}$$

$$\sum_{n \in C \cup P} \sum_{i \in I} Z_{ijn} \leq \sum_{l,m \in m; m \neq 0} G_{jm} X_{ilm}, \forall j \in W \tag{8}$$

$$\sum_{j \in W} Z_{ijp} \geq PR_{ip}, \forall i \in I, p \in P \tag{9}$$

$$Y_{isj} \leq \alpha_{ij} \cdot Y_{isj}, \forall i \in I, j \in W, s \in S \tag{10}$$

$$X_{jlm} \in (0, 1), \forall j \in W, l, m \in M \tag{11}$$

$$Y_{isj} \geq 0, \forall i \in I, s \in S, j \in W \tag{12}$$

$$Z_{ijn} \geq 0, \forall i \in I, j \in W, n \in C \cup P \tag{13}$$

In the above formulation, the objective functions maximize the percentage of satisfied demand of the general customers, while minimizing the integration cost Cost_I and the operation cost Cost_O. The integration cost Cost_I consists of the costs of closing former warehouses, and adjusting the capacities of existing warehouses. The operation cost Cost_O of the new value chain network includes the annual fixed cost of warehouse operation, as well as the variable transportation cost from the first layer to the third layer through warehouses. The percentage of satisfied demand PDC is the weighted sum of each general customer's satisfied percentage. Constraint (5) ensures that at most one capacity level is assigned to each warehouse. Constraint(6) makes the incoming and outgoing flows equal at each warehouse. Constraint (7) guarantees that the item quantity transported between the first layer and the second layer does not exceed the suppliers' production capacity. Constraint (8) guarantees that the item quantity transported between second layer and the third layer does not exceed the warehouses' capacities. Constraint (9) ensures all the requirement of the assembly plants must be satisfied. Constraint (10) ensures the appointed storage relationship. Constraint (11) ensures the integrality restriction on the variables, whereas Constraint (12)/(13) enforce the non-negativity restriction on all other decision variables.

5 Engineering Value Chain Configuration Optimization Algorithm

Several conflicting objective functions must be simultaneously optimized in the multi-objective problems. Due to the conflicting nature of the objectives, it is impossible to achieve an ideal solution in which each objective obtains its optimal value. To that end, Swiss economist Pareto introduced the concept of Pareto-optimality for multi-objective optimization. A solution of a multi-objective optimization problem is Pareto-optimal if and only if it is impossible to make at least one objective better without making anyone else worse. The set of Pareto-optimal solutions of a multi-objective optimization problem consists of all decision vectors for which the corresponding objective vectors cannot be improved in a given dimension without worsening another, and the set of all the corresponding objective vectors is called the Pareto-front. The objective of multi-objective optimization is to find its Pareto-optimal solution set and corresponding Pareto-front.

In the literature, many researchers have successfully applied genetic algorithms (GAs) to solve value chain optimization problems. As a population-based approach, genetic algorithm (GA) is appropriate in solving complex multi-objective problems. Many multi-objective GAs have been proposed, and NSGAII (Deb 2002) is one of the best multi-objective genetic algorithms as evident from the existing literature. Thus in this study, NSGAII algorithm is adopted to solve the value chain network integration model, and the detailed steps of NSGAII are shown in Fig. 3.

The whole process is repeated until the stopping criterion is met and the individuals with rank 1 in the last generation are reported. The termination criterion is usually characterized by the number of generations. The output of the algorithm is a set of non-dominated Pareto-optimal solutions, as all the solutions are the best in a sense of multi-objective optimization. In the following account, the key components of the proposed multi-objective solution procedure will be described in detail.

5.1 Problem Representation and Initialization

Choosing a good representation scheme for the construction of the genotype is one of the critical issues in using GA to solve optimization problems. In the value chain network integration problem, the network handles n types of items and each type of item has an independent two-stage logistics network (from the first-layer facilities to the second-layer facilities and from the second-layer facilities to the customer and plants). The whole value chain network can be seen as a superposition of the logistic networks of each type of item. Since a single-stage logistics network can be represented by a spanning tree (Gen 2000) and be encoded using a determinant code (Abuali 1995). We can encode the solutions of the value chain network integration problem based on tree structure. The chromosome structure is designed as shown in Fig. 4.

As shown in Fig. 4, the chromosome includes two parts, the first part represents the DC capacity levels with M integers between 0 and m, where "0" means the warehouse is closed. The second part corresponds to the logistics networks of K items. Each item has two determinant codes, representing the spanning tree of the first and second stages, respectively. The determinant code D_T for a spanning tree $T\ (V,\ E)$ is defined as follows (Abuali 1995):

$$\mathrm{D}_T = (x_2, x_3, \cdots, x_n) | (x_i, i) \in E, \mathit{for}\ i{=}2\ \mathit{to}\, n$$

where x_i is an integer between 1 and n.

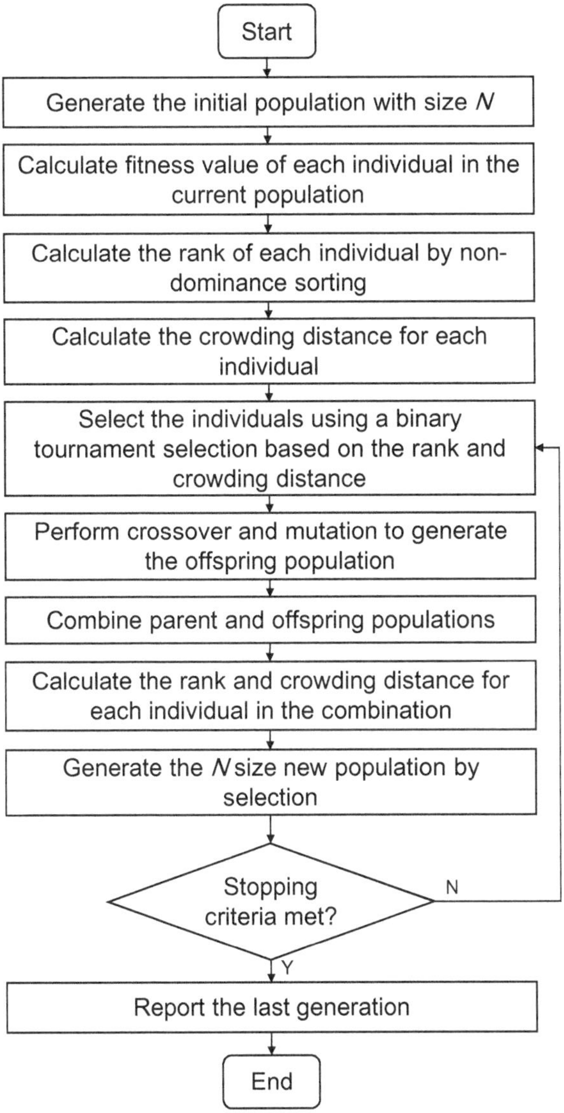

Fig. 3 NSGAII-based solution procedure

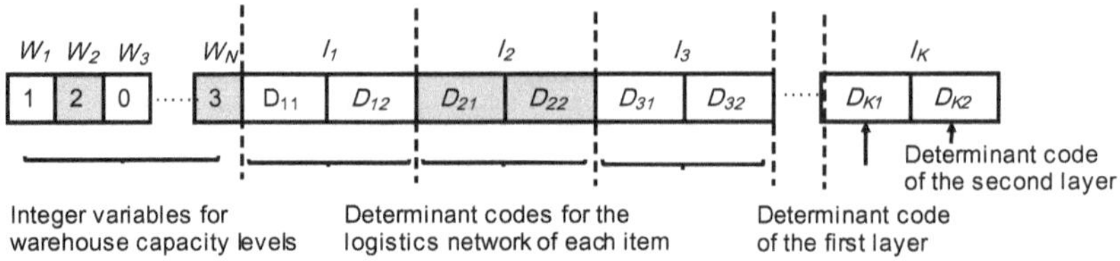

Fig. 4 Representation of chromosome

D_T has n-1 digits, and the i-1 position in D_T represents the direct connecting node of node i. For example, the determinant code for the spanning tree in Fig. 5 is (6 6 9 1 2 2 3 1). The first digit "6" means that Nodes 2 and 6 are connected in the spanning tree. The decoding flow chart of the determinant code is shown in Fig. 6.

Using the structure shown in Fig. 4, each chromosome corresponds to a value chain network and the related logistics relationships. All the decision variables can be achieved by decoding the chromosome. The decoding algorithm flow chart is shown in Fig. 7.

As previously defined, the chromosome has (1 + K) parts. The first substring consists of N integers representing the capacity levels of the warehouses. The last K substrings are determinant codes representing two-stage logistics network of K items. When decoding the chromosome,

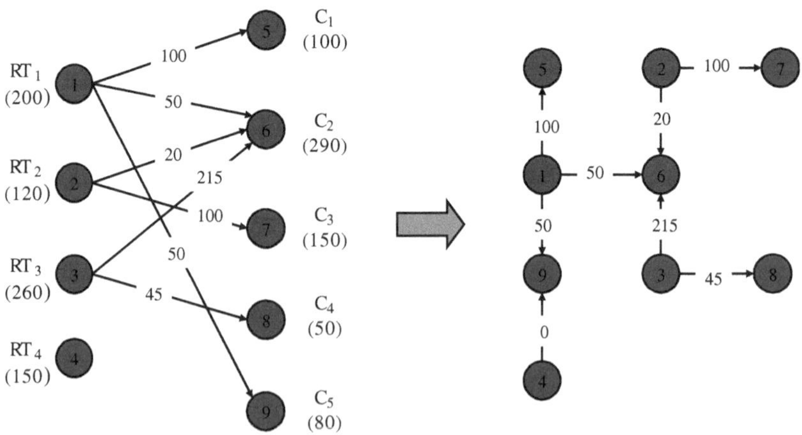

Fig. 5 Spanning tree of a single-stage logistics network

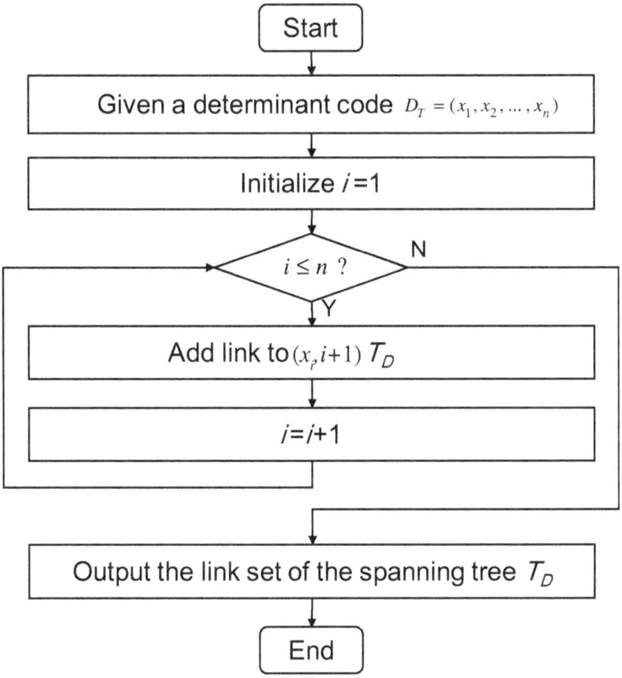

Fig. 6 Determinant decoding flow chart

the first N digitals are decoded firstly to achieve the capacity of each warehouse (if the capacity is "0" the warehouse is not included in the network). Then, the logistics network of each item is decoded successively. The chromosome substring of each item consists of two determinant codes, namely D_{k1} and D_{k2}. D_{k1} corresponds to the first stage, whereas D_{k2} corresponds to the second stage. For a certain item, the decoding process is backwards and can be divided into three steps: (1) decode D_{k2} firstly to achieve the subnetwork and supply amounts between the second layer and the third layer; (2) decode D_{k1} to achieve the subnetwork and supply amounts between the first layer and the second layer; and (3) check whether the equal flows constraints are satisfied and otherwise rearrange the supply amounts between the second layer and the third layer. In step (1), the determinant code D_{k2} is decoded first to achieve the supply links between the warehouse and the

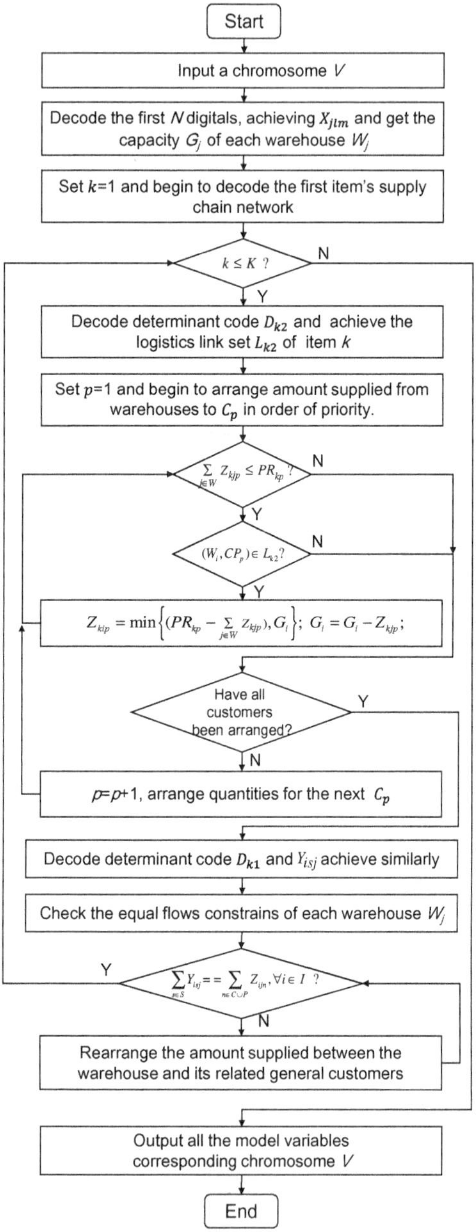

Fig. 7 Decoding algorithm

customers (including general customer and assembly plants). Due to the capacity constraints, the warehouses may not satisfy or the customers; however, according to the constraints that all the requirement of the assembly plants must be satisfied, thus the assembly plants are satisfied prior to the general customer, and furthermore, the general customers are also queued in a priority-descending order. In step (3), the rearrangements also follow the same allocation principle and the chromosome that cannot satisfy the assembly plants' requirements will be abandoned. With the above three steps, the value chain network corresponds to a certain item and the related variables are achieved. Conducting the decoding procedure on the items one by one, all the single-item value chain networks will be achieved and then the whole value chain network can be determined.

5.2 Crossover and Mutation

Based on the encoding structure, we employ two-cut-point crossover and segment-based mutation operations (Altiparmak 2009). Figure 8 shows a simple illustration of the crossover operation. Two random positions are generated as the head and tail, respectively, and alleles from the head to the tail are exchanged between parents.

A mutation is then applied to the generated offspring according to a mutation rate. As previously defined, the chromosome can be divided

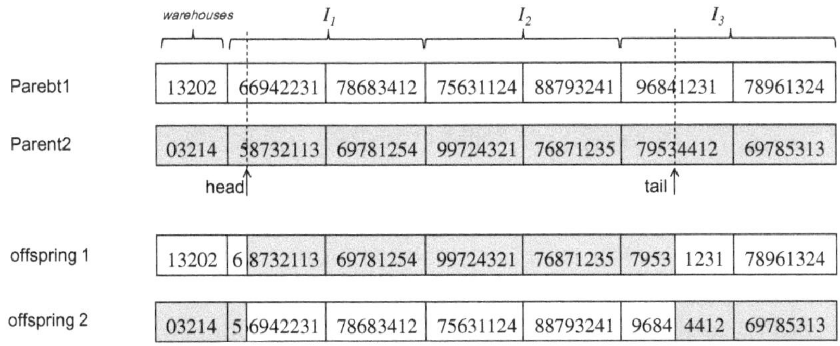

Fig. 8 Illustration of crossover operator

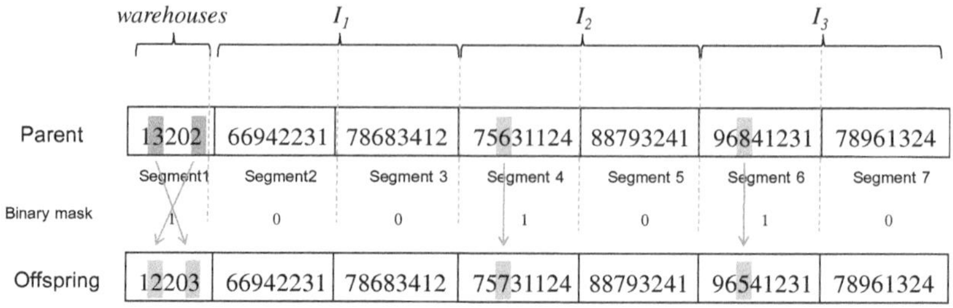

Fig. 9 Illustration of mutation operation

into (1 + K) segments, where K is the number of items. The first segments are integers representing the capacity levels of each warehouse. The last K segments are determinant codes representing the logistics network of K items. The mutation is based on the chromosome segments. Figure 9 illustrates the mutation operation.

The binary mask indicates the segments to be mutated. If the first two segments are selected, an exchange mutation operator is employed within the segment. If the last K segments are selected, the value of a randomly selected gene is replaced with a new one, ranging from one to the number of corresponding nodes.

5.3 Repair Mechanism

During determinant codes generation, chromosome crossover and mutation, infeasible codes that cannot be adapted to generate the transportation tree may be obtained. Thus, a repair mechanism is needed in the whole evolution process to check and repair the illegal determinant codes.

A spanning tree may be illegal due to three reasons: "missing Node 1"; "cycle"; and "self-loop" (Abuali 1995) ;(Chou 2001). "Cycle" and "self-loop" can be avoided in the initialization through a restriction on the encoding range of genes: in concrete terms, suppliers are numbered from 1 to M and demanders are numbered from M + 1 to M + N. According to the determinant decoding method, nodes corresponding to the first (M−1) genes connect to nodes indexed from 2 to M, which are supplier nodes. Therefore, the first ($M-1$) genes must correspond to the demander nodes, and the value range of the first (M − 1) genes is restricted to [M + 1, M + N]. Similarly, the value range of the last (N + 1) genes is restricted to [1, M], ensuring connection of demand nodes to supplier nodes. Such restrictions on the value range effectively prevent "self-loops" and "cycles".

Thus, only "missing Node 1" may occur in our problem, and the repair mechanism is as follows: Given a determinant encoding substring, an initial scan determines if Node 1 is missing. If yes, find the lowest-cost node connected to Node "1" based on the cost matrix, and assign 1 to the corresponding position. In case of a tie, randomly select a position. After this repair procedure, all the determinant codes can be decoded into spanning trees, and the distribution patterns in each stage can be determined.

6 Numerical Example and Discussion

The proposed model and solution procedure are tested with a network integration problem with three types of items: 100 first-layer facilities, 15 warehouse and 50 end customers. The algorithm was implemented using MATLAB programming language and was executed on a Pentium Dual E2200 processor (2.2 GHz clock) with 2 GB memory. The parameters were determined as: population size = 300; maximum number of generations = 300; crossover probability = 0.7; mutation probability = 0.3. Figure 10 shows the Pareto-front, and Table 1 shows part of the Pareto-optimal solutions.

The Pareto-optimal solutions represent the trade-off among the three objectives, validating the fact that the three objectives are incompatible, and that a perfect value chain network is unreachable. For example, achieving higher customer service level would require higher operation cost. Result No.1 shows that more cost is incurred to ensure that all general customer demands are satisfied. However, the general customer demand quantities are predicted based on the history of business data and market research; thus, increasing the investment to reach customer service level "1" is unadvisable. Particularly, result No.10–12 show that

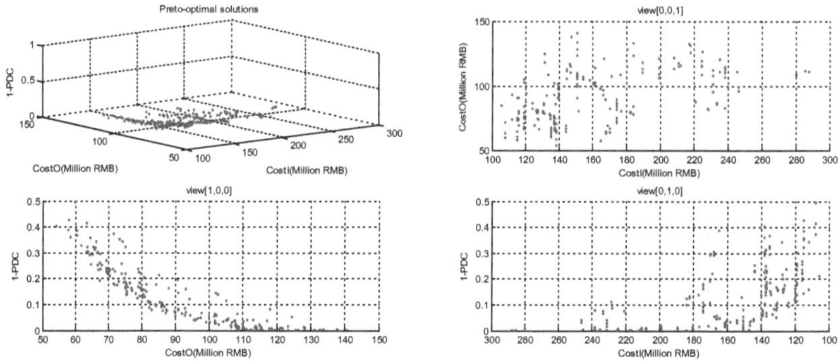

Fig. 10 The Pareto-front of the value chain network construction problem

Table 1 Part of the Pareto-optimal solutions

No.	*PDC*	$Cost_I$	$Cost_o$	*First part of the chromosome*
1	1.0000	117.35	164.52	121101020002200
2	0.9796	93.22	149.45	021100202002200
3	0.9573	85.17	134.32	220100221002200
4	0.9443	74.34	156.23	121100201002200
5	0.9339	81.88	130.34	122100220002200
6	0.9322	82.34	125.60	120100220002200
7	0.9262	78.95	135.72	202000220002200
8	0.9250	75.71	141.89	022100201002200
9	0.9004	65.33	149.24	120100221002200
10	1.0000	0	382.78	1211211111211111
11	0.9321	0	317.53	1211211111211111
12	0.9145	0	300.21	1211211111211111

to achieve the same standard service level, the operation cost ($Cost_O$) will be very high if the two networks are not integrated and optimization ($Cost_I = 0$).

7 Summary

Decision-making is a very important part of engineering value chain management. In this chapter, four popular decision-making problems (partner selection, location-allocation, inventory decision, and vehicle routing) and three decision-making methods (multi-criteria decision aid method, data mining technology-based method, and optimization model-based method) are introduced firstly. Then, the optimization model-based methods are emphasized. To interpret the decision process using optimization model-based method in detail, a complex value network integration and optimization problem is investigated as an example, a multi-objective optimization model is developed, and the genetic algorithm-based solution procedure is proposed to achieve the network integration and optimization solutions. Finally, numerical experiment and discussion are conducted to demonstrate the benefit of integration and optimization. The decision example shows the effectiveness and advances of solving management problems by information technologies. How to formulate a realistic decision problem and develop appropriate optimization algorithm are the crux and difficulty of the optimization model-based decision method. A stochastic or fuzzy model would be more realistic, but increasingly complex. Future work can focus on formulating the stochastic model and developing an effective solution algorithm.

References

Abuali, F. N., Wainwright, R. L., & Schoenefeld, D. A. (1995, July). Determinant factorization: A new encoding scheme for spanning trees applied to the probabilistic minimum spanning tree problem. *Proceedings of the Sixth International Conference on Genetic Algorithms*, (pp. 470–477).

Altiparmak, F., Gen, M., Lin, L., & Karaoglan, I. (2009). A steady-state genetic algorithm for multi-product supply chain network design. *Computers & Industrial Engineering, 56*(2), 521–537.

Bakker, M., Riezebos, J., & Teunter, R. H. (2012). Review of inventory systems with deterioration since 2001. *European Journal of Operational Research, 221* (2), 275–284.

Başligil, H., Kara, S. S., Alcan, P., Özkan, B., & Çağlar, E. G. (2011). A distribution network optimization problem for third party logistics service providers. *Expert Systems with Applications, 38*(10), 12730–12738.

Beamish, P. W. (1987). Joint ventures in LDCs: Partner selection and performance. *Management International Review*, 23–37.

Boer, L. D., Labro, E., & Morlacchi, P. (2001). A review of methods supporting supplier selection. *European Journal of Purchasing & Supply Management, 7* (2), 75–89.

Boyd, S., & Vandenberghe, L. (2004). *Convex optimization*. Cambridge: Cambridge University Press.

Chai, J., Liu, J. N., & Ngai, E. W. (2013). Application of decision-making techniques in supplier selection: A systematic review of literature. *Expert Systems with Applications, 40*(10), 3872–3885.

Charnes, A., & Cooper, W. W. (1957). Management models and industrial applications of linear programming. *Management Science, 4*(1), 38–91.

Chou, H., Premkumar, G., & Chu, C. H. (2001). Genetic algorithms for communications network design-an empirical study of the factors that influence performance. *IEEE Transactions on Evolutionary Computation, 5*(3), 236–249.

Cooper, L. (1963). Location-allocation problems. *Operations Research, 11*(3), 331–343.

Creazza, A., Dallari, F., & Rossi, T. (2012). Applying an integrated logistics network design and optimisation model: The pirelli tyre case. *International Journal of Production Research, 50*(11), 3021–3038.

Dantzig, G. B., & Ramser, J. H. (1959). The truck dispatching problem. *Management Science, 6*(1), 80–91.

Deb, K., Pratap, A., Agarwal, S., & Meyarivan, T. (2002). A fast and elitist multiobjective genetic algorithm: NSGA-II. *IEEE Transactions on Evolutionary Computation, 6*(2), 182–197.

Eksioglu, B., Vural, A. V., & Reisman, A. (2009). The vehicle routing problem: A taxonomic review. *Computers & Industrial Engineering, 57*(4), 1472–1483.

Farahani, R. Z., Asgari, N., Heidari, N., Hosseininia, M., & Goh, M. (2012). Covering problems in facility location: A review. *Computers & Industrial Engineering, 62*(1), 368–407.

Fonseca, C. M., & Fleming, P. J. (1993, June). Genetic algorithms for multiobjective optimization: Formulation discussion and generalization. *Proceedings of the Fifth International Conference on Genetic Algorithms,* (pp. 416–423).

Gen, M., & Cheng, R. (2000). *Genetic algorithms and engineering optimization*, Vol. 7. Wiley.

Govindan, K., Rajendran, S., Sarkis, J., & Murugesan, P. (2015). Multi criteria decision making approaches for green supplier evaluation and selection: A literature review. *Journal of Cleaner Production, 98,* 66–83.

Gumus, A. T., Guneri, A. F., & Keles, S. (2009). Supply chain network design using an integrated neuro-fuzzy and MILP approach: A comparative design study. *Expert Systems with Applications, 36*(10), 12570–12577.

Horenbeek, A. V., Buré, J., Cattrysse, D., Pintelon, L., & Vansteenwegen, P. (2013). Joint maintenance and inventory optimization systems: A review. *International Journal of Production Economics, 143*(2), 499–508.

Horn, J., Nafpliotis, N., & Goldberg, D. (1994, June). A niched pareto genetic algorithm for multiobjective optimization.In *Proceedings of the First IEEE Conference on Evolutionary Computation. IEEE World Congress on Computational Intelligence*, (pp. 82–87).

Kiresuk, T. J., & Sherman, M. R. E. (1968). Goal attainment scaling: A general method for evaluating comprehensive community mental health programs. *Community Mental Health Journal, 4*(6), 443–453.

Knowles, J., & Corne, D. (1999, July). The pareto archived evolution strategy: A new baseline algorithm for paretomultiobjective optimisation. In *Proceedings of the 1999 Congress on Evolutionary Computation.*

Lee, J. H. (2014). Energy supply planning and supply chain optimization under uncertainty. *Journal of Process Control, 24*(2), 323–331.

Lin, C., Choy, K. L., Ho, G. T., Chung, S. H., & Lam, H. Y. (2014). Survey of green vehicle routing problem: Past and future trends. *Expert Systems with Applications, 41*(4), 1118–1138.

Melo, M. T., Nickel, S., & Saldanha-da-Gama, F. (2009). Facility location and supply chain management—A review. *European Journal of Operational Research, 196*(2), 401–412.

Monteiro, M. M., Leal, J. E., & Raupp, F. M. P. (2010). A four-type decision-variable MINLP model for a supply chain network design. *Mathematical Problems in Engineering, 2010.*

Nemhauser, G. L. (1989). *Handbook in Operations Research and Management Science.* North-Holland.

Pillac, V., Gendreau, M., Guéret, C., & Medaglia, A. L. (2013). A review of dynamic vehicle routing problems. *European Journal of Operational Research, 225*(1), 1–11.

Schaffer, J. D. (1985, July). Multiple objective optimization with vector evaluated genetic algorithms. In *Proceedings of the First International Conference on Genetic Algorithms* , (pp. 93–100).

Snyder, L. V. (2006). Facility location under uncertainty: A review. *IIE Transactions, 38*(7), 547–564.
Srinivas, N., & Deb, K. (1994). Muiltiobjective optimization using nondominated sorting in genetic algorithms. *Evolutionary Computation, 2*(3), 221–248.
Turban, E., & Meredith, J. R. (1998). *Fundamentals of management science.*New york: McGraw-Hill College.
Zitzler, E., & Thiele, L. (1999). Multiobjective evolutionary algorithms: A comparative case study and the strength Pareto approach. *IEEE Transactions on Evolutionary Computation, 3*(4), 257–271.

Authors' Biography

Lina Zhou is a Lecturer at the School of Computer Science and Technology of Harbin Institute of Technology, China. She has extensive experience in industry information management system development. Her research interests include supply chain management and optimisation, service engineering, machine learning and intelligent decision-making.

Xiaofei Xu is professor of Computer Science and Technology at Harbin Institute of Technology, China. He is President of Harbin Institute of Technology (Weihai). He chairs the Technical Committee on Service Computing of China Computer Federation.

10

Engineering Value Chain Coordination and Optimization

Ming Dong and Yonglin Li

1 Introduction

Increasing competitive pressures and market globalization are forcing firms to develop engineering chain that can quickly respond to customer needs. Historically, have been managed independently, buffered by large inventories. Therefore, effective engineering management requires coordination among the three fundamental stages of the engineering chain: procurement, production, and distribution.

Coordination can be visualized in different functions such as logistics, inventory management, forecasting, and transportation. Similarly, various interfaces such as supplier—manufacturer and manufacturer—retailer can be effectively managed using coordination. The members of engineering chain are often separate and independent economic entities. Even though

M. Dong (✉) · Y. Li
Shanghai Jiao Tong University, Shanghai, China
e-mail: mdong@sjtu.edu.cn

Y. Li
e-mail: liyonglin7053@qq.com

Y. Zhang and M. Gregory (eds.), *Value Creation through Engineering Excellence*,
https://doi.org/10.1007/978-3-319-56336-7_10

coordination improves the performance of the engineering chain, it may not always be beneficial to coordinate the engineering chain members. Hence, a key issue in engineering chain coordination is then to develop specific mechanisms that align the objective of independent members and coordination, their decisions, and activities so as to optimize engineering system performance. By utilizing coordination mechanisms, the performance of engineering chain value may improve. There are four different types of coordination mechanisms as discussed (see Fig. 1).

With the advances in logistics and engineering chain management technology in recent years, there has been an explosion of interest in the topic of "engineering Chain Optimization". Optimization is the application of processes and tools to ensure the optimal operation of manufacturing and distribution chain. This often involves the application of mathematical modeling techniques using optimization techniques (see Fig. 2).

Typically, engineering chain optimization is trying to maximize the profitable operation of their manufacturing and distribution chain. This could include measures like maximizing gross margin return on inventory invested (balancing the cost of inventory at all points in the engineering chain with availability to the customer), minimizing total operating expenses (transportation, inventory, and manufacturing), or maximizing the gross profit of products distributed through the

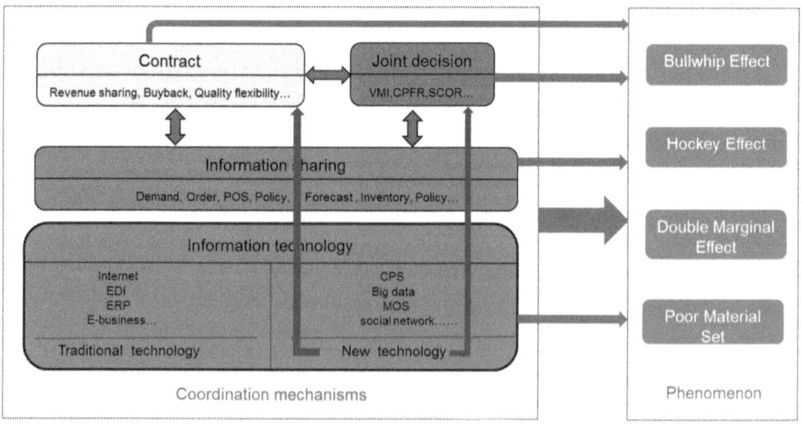

Fig. 1 The relationships of coordination mechanisms

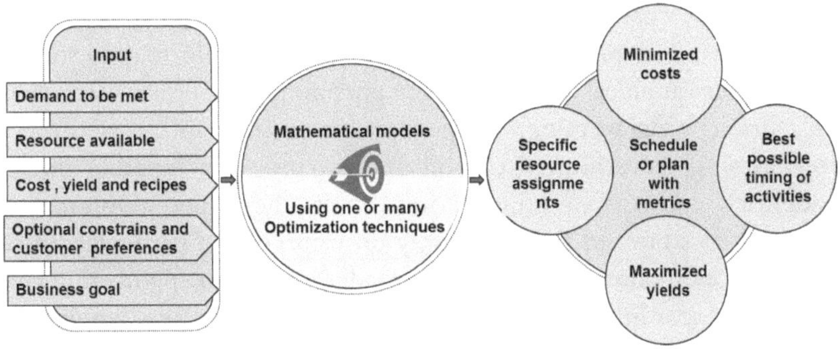

Fig. 2 The processor of modeling and optimization in EV chain

engineering chain. Engineering chain optimization involved the end-to-end process, which starts with the design of the product or service and ends with the time when it has been sold, consumed, and finally, discarded by the consumer; issues can be classified into two broad categories: configuration (design-oriented) issues that relate to the basic infrastructure on which the engineering chain executes and coordination (execution-oriented) issues that relate to the actual execution of the engineering chain.

The competitiveness and dynamic nature of today's marketplace are due to rapid advances in information technology, short product life cycles and the continuing trend in global outsourcing. Managing the resulting engineering chain networks effectively is a complex and challenging task which is imputable to a high level of uncertainty in engineering demand, conflict objectives, and vagueness of information, numerous decision variables, and constraints. System modeling is used in such cases to model the real system. These models can be mathematical models or simulation models. In order to capture the system complexity, mathematical models are rarely used to model engineering chain. Decisions are made at different levels in engineering chain. These decisions are needed to be supported by robust optimization techniques to enable decisions to evaluate the impact of their decisions prior to actually making them in the real environment.

Optimization techniques have shown a great potential to solve engineering chain problems that cause an immense challenge to decision makers. These challenges are imputable to a high level of uncertainty in supply demand, conflict objectives, lack of needed information, numerous decision variables, and inevitable constraints. Traditional techniques (e.g., linear programming, integer programming, and mixed-integer programming) have limited capabilities to handle the inherent interdependencies in current engineering chain networks. Optimization methodologies have to focus not only on improving a particular process performance but also on achieving a broader impact on engineering chain efficiency. Accordingly, metaheuristics were presented into engineering chain applications because of its global optimization capabilities in stochastic environments. On the other hand, statistical methods and metamodel-based methods can be incorporated with metaheuristics to provide more reliable solutions in a reasonable timeframe.

Manufacturing processes are also undergoing major challenges to achieve the Smart Factory vision such as to increase systematic processes, reuse, and improve understandability of complex structures. Most of the new factory concepts share attributes of smart networking (Dolgui and Proth 2010; Ivanov et al. 2013; Chick et al. 2014). That is why it becomes a timely and crucial topic to consider engineering chain as collaborative cyber-physical systems (Camarinha-Matos and Macedo 2010; Ivanov et al. 2014). Cyber-physical systems incorporate elements from both information and material subsystems which are integrated and decisions in them are cohesive (Zhuge 2011).

Some coordination mechanisms, such as quantity discount schedule and revenue-sharing schedule, are used to regulate the relationship among ``'s members. The continuous evolving dynamic structure of the engineering chain poses many interesting challenges for effective system coordination. Very often, schedules are designed for the static environment such as a known market demand and a distribution function in the stochastic environment. These schedules can be defined as a static coordination mechanism. However, after the plan has been settled down, the environment is often disrupted by some unexpected events, such as machine breakdown, the raw material shortage, the SARS epidemic, and Hurricane Katrina. The disruptions have made companies aware of the need for active disruption management.

How to design adaptive coordination mechanisms is a problem, and some aspects can be considered as follows:

- Double check the traditional coordination mechanisms by further consideration of the more realistic business environment. For example, considering the influence of the widely used trade credit and its associated risk in realistic commerce.
- Realize the influences of a new relationship of customs and behaviors. For example, based on a review of the social network, consumption habits, and the ways of communication.
- Realize the influences of new ICT. For example, forecast demand directly using big data mining technology.
- Holistic coordination model for integrating currently commercial environment and novel manufacturing models.
- It must be noted that a typical engineering chain also deals with human systems, and hence, it is hard to coordinate engineering chain members may be visualized.
- There exist differences in the interest of engineering chain members as the members habitually work as an individual based on local perspective and opportunistic behavior results in a mismatch of supply and demand.

Today's markets call for elaborate competition schemes as a large variety of products is available to meet customer requirements. Mass customization has become an imperative for many manufacturers to survive in the growing competition characterized by heterogenic customer demands, accelerated new product development investments, and shortened product life cycles. Duray shows that the degree or type of customization depends on the point in the production cycle where the initial customer involvement is (Duray et al. 2000). They define four points in the production cycle, where each of the points is an expression of the degree of customization: design, fabrication, assembly, and use. The first three are quite easily recognized as commonly known, engineer-to-order (ETO), make-to-order (MTO), and assembly-to-order (ATO).

In summary, engineering chain management in new business environment is complicated and has its own special characters mainly reflected in two aspects. One is the random information from customer orders and the complex relations among engineering chain cooperators, which can cause many complicated contradictions in strategic or operational level and bring dynamic or stochastic characteristics to it; the other is the outstanding relations of collaborative benefits and risks in this complicated environment. Therefore, we must probe the ways to respond to these characteristics and analyses, the coordination mechanisms, and optimization approaches in new business environment.

Coordination is essential for successful engineering chain management; we make a conjoint research of information sharing, operational research and behavior research to adaptive coordination mechanism, and the details are worded in Sect. 2 . As for modeling and optimization, review papers (Da Silveira et al. 2001; Fogliatto et al. 2012) defined four steps that described activities in generating and processing MC orders, namely (1) building the product catalog, (2) configuring customer orders, (3) transferring orders to manufacturing, and (4) manufacturing customized orders. We focus on step (3) to engineering chain optimization, transferring orders to manufacturing, which is specific to the manufacturer who fulfills the orders based on available production resources: materials and production capacity under the ATO engineering chain environment in Sect. 3.

2 Trust-Embedded Coordination in Information Sharing

2.1 Introduction

Information sharing is one of the most important coordination mechanisms in engineering chain (see Fig. 1). In industries, many firms follow electronic data interchange (EDI) system to place orders. Based on the received orders, their upper streamers determine their optimal capacities (Premkumar et al. 1994). Because the orders are costless, non-verifiable, and cancelable before shipping, they are commonly referred to soft-orders (Taylor and Plambeck 2007). Therefore, EDI based non-binding soft-orders primarily do not involve complex contracts. Since a downstream retailer has an incentive to over-order products for abundant supply, fully relying on the soft-orders usually leads to great capacity risks. Sometimes, an upstream supplier deems all soft-orders to be meaningless (Cai et al. 2013). Based on the above analysis, a critical question of information sharing by soft-orders is how much information to be credibly transmitted and what is the optimal decision under information asymmetric circumstance.

Information asymmetry in a supplier—retailer relationship is well studied; strategic information transmission and contract designing are helpful to align the pecuniary incentives of engineering chain partners and ensure credible information sharing (e.g., Cachon and Lariviere 2001; Özer et al. 2011). In industries, many firms share nonbinding unverified information via soft-orders. For example, Nike has started soft-order service via an EDI system for nearly 15 years. At the first year when the service was started, the service helped Nike to reduce its stock by 14% (Nike Inc. 1999). It is also reported that more than 41% of Asia Pacific firms have adopted EDI systems to transmit their soft-orders since 2008. Motivated by the practices, some behavior studies (e.g., Özer et al. 2011; Ebrahim-Khanjari et al. 2012) establish the role of trust in information sharing via soft-orders without complex contracts and verify that trust is the primary factor for credible information transmission. Since "trust is a psychological state comprising the intention to accept vulnerability based upon positive expectations of the intentions or behavior of another" (Rousseau et al. 1998), it is also affected by instant behaviors in current transactions. In order to mathematically present the influence of instant behaviors to trust, we suggest a trust evaluation model. The proposed model is helpful to analyze the value of trust in the information sharing process and how trust affects engineering chain decisions.

We focus on a two-tier engineering chain consisted of a supplier and a retailer. Both the supplier and retailer do forecast independently before transaction. At the beginning of their transaction, the retailer places soft-orders and the supplier decides his capacity afterward. Because both supplier and retailer's forecasted demands are their private information, it is risky for both sides to make decisions under the highly information asymmetric circumstance. For example, the retailer does not know how much the supplier trusts when she places orders because the supplier's forecasted demand is unknown to her. Simultaneously, the supplier also faces potential loss stemming from the retailer's artificial soft-orders since he does not know the retailer's private forecast. Therefore, we provide a trust-embedded coordination to mitigate their risks and improve engineering chain performances. The coordination process consists of two stages: at the first stage, the retailer and supplier negotiate a cost-sharing rule; at the second stage, the retailer makes ordering decision and the supplier makes capacity decisions sequentially. We are interested in: what is the role of trust takes in their transaction? How does the negotiation power of the supplier/retailer affect the engineering chain decisions and performances? Whether the proposed trust-embedded coordination works effectively or not?

2.2 Modeling Trust

The existing psychology theories have proven that trust is affected by multiple factors. Some factors can be evaluated before a transaction, e.g., reputation, historical transaction, and peer recommendations from trustees; emotions, experience, and cognition from the trustors. The evaluations of these factors are pre-known and unchanging in a transaction, so that these factors can be named as predetermined factors. However, some instant behaviors in transactions can affect trust (Rousseau et al. 1998). Thus, some trust-affecting factors (i.e., instant behaviors) can only be evaluated in the current transaction. Thus, we name the factors that work in the current transaction as instant factors. For convenience, we denote all the predetermined factors by R and all instant factors by Δ. Moreover, because trust is a kind of psychological state, its distributions are often evaluated by regression approaches (Laeequddin et al. 2012). Thus, we consider trust T as a randomly distributed variable with cumulative density function (c.d.f) $F(t)$ and probability density function (p.d.f.) $f(t)$. According to the above analysis, trust T is formulated as:

$$T \sim f(\mathrm{t}/\mathrm{R}, \Delta) \quad \text{where} \quad 0 \leq t \leq 1 \tag{1}$$

Equation (1) suggests a general model to quantify trust, where R denotes predetermined factors and Δ denotes instant factors. Since R and Δ might contain different items in different situations, the proposed trust model is general and applicable to different engineering chain problems. For example, a decision maker might analyze information in some complex situations, e.g., multiple partners, multiple engineering chain tiers, or complex transaction processes. The decision maker can classify all trust-affecting factors of his/her problem into two groups, the predetermined factors R and the instant factors Δ. Based on this classification, his/her trust level can be evaluated by Eq. (1). Therefore, our proposed trust evaluation model is general and could be used in the complex situations.

In the target problem of this paper, the predetermined trust-affecting factors include reputation and historical transactions of the retailer, experience and psychology state of the supplier, peer recommendation from the third side, and so on. When a transaction begins, the retailer places a soft-order. After that, the supplier calculates the information mismatch level by comparing his private forecasting demand with the retailer's soft-order. Because information mismatch harms trust level l (Kosfeld et al. 2005; Sriram 2005), the supplier updates his trust based on information mismatch. We let μ_S be the supplier's forecasted demand and μ_{RS} be the retailer ordered quantity; thus, the information mismatch can be denoted by $\Delta = \frac{|\mu_{RS} - \mu_S|}{\mu_s}$. Therefore, when $\Delta = 0$, we have $T \sim f(\mathrm{t}|\mathrm{R}, 0)$, which means the supplier's trust only depends on the predetermined factor R. Thus, the supplier's trust when $\Delta = 0$ can be named as "initial trust." As suggested by Özer et al. (2011), we have trust level T ranged within [0,1]. The fact of $T = 0$ suggests that the supplier fully distrusts the retailer's soft-order, while the fact of $T = 1$ indicates the supplier fully trusts the retailer's soft-order.

In this context, market demand is formulated as $D = \mu_0 + \varepsilon$, where μ is a positive constant denoting average market demand and ε describes demand fluctuation. They both know that ε is a random variable with c.d.f. $\Gamma(\varepsilon)$ and p.d.f. $\tau(\varepsilon)$. In the engineering chain we studied, both the retailer and supplier forecast the value of μ_0 individually. Because the retailer is more close to consumers and professional on marketing, we assume the retailer can precisely forecast the distribution of market demand. Although the retailer's forecasted demand is μ_R, she places a soft-order μ_{RS} to the supplier. After receiving the retailer's soft-order μ_{RS}, the supplier updates his demand evaluation based on ordered quantity μ_{RS} and his own forecasted demand quantity μ_S. As suggested by Clemen and Winkler (1999), we assume that the supplier combines two demands of μ_{RS} and μ_S using a simple weighted average approach. Therefore, the supplier believes that the average market demand is:

$$\mu = T\mu_{RS} + (1 - T)\mu_S, \quad \text{where} \quad T \sim f(t|R, \Delta) \tag{2}$$

Since μ is a random variable, we let $G(t)$ and $g(t)$ denote its c.d.f. and p.d.f., respectively. Because $T \in [0, 1]$, we have $\min(\mu_{RS}, \mu_S) \leq E(\mu) \leq \max(\mu_{RS}, \mu_S)$ based on Eq. (2), which indicates that the supplier insists that average demand is within his own forecasted demand and the retailer's soft-order. Although the retailer does not know the value of μ_S, she can employ the concept of Bayes' rule to evaluate it. We assume that she evaluates μ_S to be μ'_S. Therefore, the retailer believes her trustworthiness T' to be:

$$T' \sim f(t'|R, \Delta') \quad \text{where} \quad t' \in [0, 1], \; \Delta' = \frac{|\mu_{RS} - \mu'_S|}{\mu'_S} \tag{3}$$

Let μ'_S denote the retailer's evaluation on μ_S, the retailer's evaluation of μ is written as:

$$\mu' = T'\mu_{RS} + (1 - T')\mu'_S \tag{4}$$

Since T' is a random variable, μ' is also a random variable.

2.3 Profit Functions

Because of advantages in information collection and customer demand forecasting, retailers in engineering chain most probably know the real demand information. Similarly, to some existing studies (e.g., Cachon and Lariviere 2001; Özer and Wei 2006), we assume that the retailer knows the demand information since she is more close to market. We also assume that the wholesale price p_S , retail price p_R, production cost c, and capacity preparation cost c_k for each product are known.

Both the supplier and retailer have their individual forecasts and communicate with soft-orders; their transaction follows such a sequence: (1) both the retailer and supplier forecast the average market demand and the retailer evaluates the supplier's forecasted demand; (2) the retailer places a soft-order to the supplier; (3) the supplier evaluates the market demand based on the retailer's soft-order and his private forecast. After that, he determines his optimal production capacity. The decision process is illustrated in Fig. 3.

Different from the decision process given by Özer et al. (2011), we do not only analyze the supplier's optimal determinations but also solve the retailer's optimal decision in this paper. As shown in Fig. 1, the engineering chain decisions are made following a Stackelberg game, where the retailer is the leader and the supplier is the follower. In the game, the retailer decides her ordered quantity at first and the supplier determines his optimal capacity based on the retailer's decision.

1. The retailer's decision.

Let Q denote the supplier's capacity, the retailer's profit is:

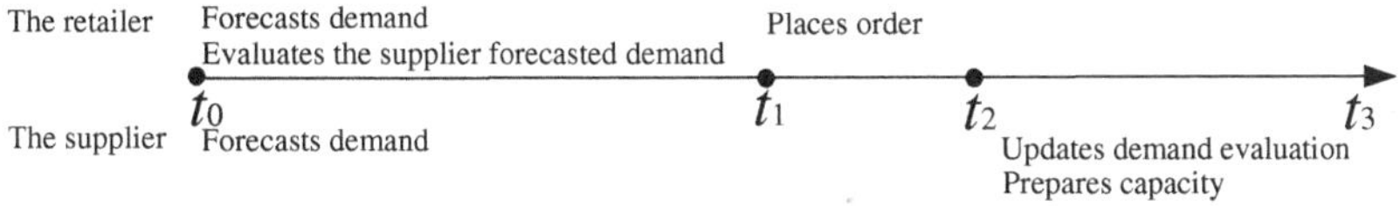

Fig. 3 Sequence of events in decentralized pattern

$$\Pi_R = \underset{\varepsilon}{E}[(p_R - p_S)\min(Q, \mu_R) + \varepsilon)] \tag{5}$$

Because the retailer evaluates μ_S to be μ'_S (Eq. 4), the supplier's expected profit in retailer's belief is:

$$\Pi'_S = \underset{\varepsilon}{E}[(p_S - c)\min(Q, \mu'_S + \varepsilon) - c_k Q] \tag{6}$$

Note that formula $\mu'_S + \varepsilon$ is random, we assume its c.d.f. and p.d.f. to be $G'(t')$ and $g'(t')$, respectively. Let μ^*_{RS} and Q'^* be the optimal solutions of $\max(\Pi_R)$ and $\max(\Pi'_S)$, respectively. We can solve the retailer's decision Q'^* by Eq. (6). Then, we introduce Q'^* into Eq. (5) and compute the retailer's decision μ^*_{RS} (Corollary 1).

Corollary 1 For $\forall \mu_R$, we have

$$\mu^*_{RS} = \arg\max_{\mu_{RS}} Q'^*,\ Q'^* = G'^{-1}\left(\frac{p_S - c - c_k}{p_S - c}\right) + E(\mu')$$

Proof. Because the retailer and supplier make decisions sequentially, we solve the supplier's decision at first. For a given μ_{RS}, we have $\mu' = T'\mu_{RS} + (1 - T')\mu'_S$, where $T' \sim f(t'|R, \Delta')$ (Eq. 3). We introduce μ' into Eq. (6) and have:

$$\frac{d\Pi'_S}{dQ} = (p_S - c)[1 - G'(Q - E(\mu'))] - c_k$$
$$\frac{d^2\Pi'_S}{dQdQ} = -(p_S - c)g'(Q - E(\mu')) < 0$$

Therefore, according to Eq. (1), we have the supplier's optimal solution as follows:

$$Q'^* = \arg\max_{Q} \Pi'_S = G^{-1}\left(\frac{p_S - c - c_k}{p_S - c}\right) + E(\mu')$$

Because $\prod_R$ strictly increases with Q, we have the retailer's decision on optimal soft-order:

$$\mu_{RS}^* = \arg\max_{\mu_{RS}} \Pi_R = \arg\max_{\mu_{RS}} Q'^*.$$

In Corollary 1, Q'^* is the retailer's evaluation about the supplier's capacity when she orders μ_{RS}^*. Therefore, the retailer's maximum expected profit is $\prod_R'^*(Q'^*)$.

2. The supplier's decision.
   After receiving the retailer's soft-order μ_{RS}, the supplier evaluated market demand is $\mu(\mu_{RS})$ by Eq. (2). We have the supplier's profit function:

$$\begin{aligned} \Pi_S &= \mathop{E}_{\varepsilon}[(p_S - c)\min(Q, \mu(\mu_{RS}) + \varepsilon) - c_k Q] \\ &\Rightarrow Q^* = \arg\max_Q \Pi_S = E[T\mu_{RS} + (1-T)\mu_S] + G^{-1}\left(\frac{p_S - c - c_k}{p_S - c}\right) \end{aligned} \tag{7}$$

Introducing μ_{RS}^* into Eq. (7), we have the supplier's optimal capacity decision:

$$Q^* = E[T\mu_{RS}^* + (1-T)\mu_S + G^{-1}\left(\frac{p_S - c - c_k}{p_S - c}\right)$$

The solutions of μ_{RS}^* and Q^* indicate that both the retailer and supplier's optimal decisions are directly linked with μ_S'.

Engineering chain ineffectiveness resulting from information asymmetric is a classic and well-documented problem in information sharing studies, and it is proved that both the supplier and retailer face potential losses when they make decisions in a decentralized pattern. Therefore, we are interested in designing a trust-embedded coordination mechanism to mitigate both the partners' risks of loss and show how much the engineering chain can benefit from the coordination.

Due to the poor engineering chain performance resulting from double marginalization, many contracts are proposed to coordinate the engineering chain performances, e.g., quantity discount contract, buy-back contract, and wholesale price contract (Cachon 2003). Different from the contracts above, we consider the supplier's trust in building a cost-sharing contract.

In the contract we proposed, both the retailer and supplier negotiate a cost-sharing rule right after their forecasts/evaluations are made. The engineering chain decisions are made following a two-stage decision process (Fig. 4). At Stage 1, both the supplier and retailer negotiate a cost-sharing rule under the information asymmetric circumstance. Once they reach an agreement on shared cost of per ordered product after the negotiation, the order is regarded as "bounded." That is, when the retailer places a soft-order and pays the negotiated shared cost, the supplier guarantees the order to be fully satisfied. Note that if they cannot successfully negotiate a cost-sharing rule, then the retailer's order will not be guaranteed by the supplier, and the order is regarded as "unbounded." At Stage 2, the retailer determines her ordered quantity and the supplier decides his optimal capacity.

Stage 1. Negotiation on shared cost.

Let m be the shared cost for each ordered product in the cost-sharing contract and let $\tilde{\Pi}_R = \underset{\varepsilon}{E}[(p_R - p_S)\min(Q, \mu_R + \varepsilon)]$ where $Q \geq \mu_{RS}$, then we have the retailer's expected profit in binding pattern as $\tilde{\Pi}_R - \mu_{RS}m$. Note that the retailer shares cost to bind her order only if the profit in binding pattern is no less than that in unbinding pattern,

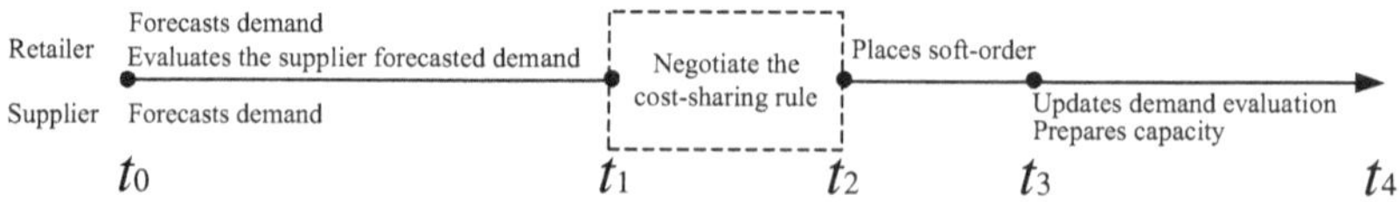

Fig. 4 Sequence of events in the coordination pattern

i.e., $\tilde{\Pi}_R^* - \mu_{RS}m \geq \Pi_R^*$. We name the constraint as "retailer's binding constraint," which can be transformed as:

$$m \leq \frac{\tilde{\Pi}_R^* - \Pi_R^*}{\mu_{RS}} \tag{8}$$

When the retailer's soft-order is bounded, the supplier's capacity is no less than the retailer's order. Thus, let $\tilde{\Pi}_S = \underset{\varepsilon}{E}[(p_S - c)\min(Q, \mu + \varepsilon) - c_k Q]$ where $Q \geq \mu_{RS}$, we have the supplier's profit in binding pattern $\tilde{\Pi}_S + \mu_{RS}m$. Simultaneously, the supplier will agree to bind retailer's soft-order only if his expected profit from binding is no less than that in unbinding pattern. For a given ordered quantity μ_{RS}, the supplier's maximum profit in unbinding pattern is calculated by $\Pi_S^* = \Pi_S[Q^*(\mu_{RS})]$ (Eq. 7). Thus, we have $\tilde{\Pi}_S^* + \mu_{RS}m \geq \Pi_S^*$. We name the constraint as "supplier's binding constraint," which can be transformed as:

$$m \geq \frac{\tilde{\Pi}_S^* - \Pi_S^*}{\mu_{RS}} \tag{9}$$

Equations (8–9) give the retailer and supplier's requirements on binding the retailer's soft-orders. Therefore, there exist two negotiation results.

Negotiation result 1: soft-order is not bounded.

Based on Eqs. (8) and (9), we have $m \in \varphi$ if $\tilde{\Pi}_S^* - \Pi_S^* > \Pi_R^* - \tilde{\Pi}_R^*$. Therefore, the soft-order is not bounded under this situation, and both the supplier and retailer make decisions in unbinding pattern.

Negotiation result 2: soft-order is bounded.

When $\tilde{\Pi}_S^* - \Pi_S^* \leq \Pi_R{}^* - \tilde{\Pi}_R{}^*$, the supplier and retailer are able to negotiate a value of shared cost m, where $m \in \left[\frac{\tilde{\Pi}_S^* - \Pi_S^*}{\mu_{RS}}, \frac{\Pi_R^* - \tilde{\Pi}_R^*}{\mu_{RS}}\right]$. Because both the supplier and retailer have their own private demand forecasts, the cost-sharing rule is negotiated depending on their negotiation power in the engineering chain. As suggested by Blodgett that the negotiation power is proportionally linked with profit-sharing, we denote w, where

$w \in (0,1)$, as the supplier's negotiation power and $1-w$ as the retailer negotiation power, respectively. Let $m_S = \frac{\tilde{\Pi}_S^* - \Pi_S^*}{\mu_{RS}}$ and $m_R = \frac{\tilde{\Pi}_R^* - \Pi_R^*}{\mu_{RS}}$, the negotiated shared cost $m^*(\mu_{RS})$ for each bounded product in the first stage is written as $m^*(\mu_{RS}) = m_S + w(m_R - m_S)$. Under the extreme situation that the retailer dominates the engineering chain (i.e., $w \to 1$), we have $m^*(\mu_{RS}) \to m_R$; otherwise, if the supplier dominates the engineering chain, we have $m^*(\mu_{RS}) \to m_S$.

Stage 2. Order and capacity decisions.

In this stage, the retailer and supplier make decisions sequentially based on the negotiation results at Stage 1.

Under negotiation result 1, the retailer and supplier make decisions in unbinding pattern. Thus, their optimal solutions can be obtained based on Corollary 1:

$$\{\mu_{RS}^*, Q^*\} = \left\{ \arg\max_{\mu_{RS}} \left[G'\left(\frac{p_S - c - c_k}{p_S - c}\right) + E(\mu')\right], E[\mu_S T(\mu_{RS}^*) + \mu_{RS}^* - \mu_{RS}^* T(\mu_{RS}^*)] + G^{-1}\left(\frac{p_S - c - c_k}{p_S - c}\right)\right\} \tag{10}$$

However, under the negotiation result 2, the retailer has her order bounded by sharing a cost $m^*(\mu_{RS})$ for per unit of ordered products. Thus, the supplier takes the retailer ordered quantity as lower bound of his capacity.

1. The retailer's decision.

Let $\Pi'_S = \underset{\varepsilon}{E}[(p_S - c)\min(Q, \mu'_S + \varepsilon) - c_k Q]$, then the supplier's expected profit from the retailer's belief is $\Pi'_S + \mu_{RS} m^*(\mu_{RS})$. Therefore, the supplier's optimal reaction Q'^* from the retailer's belief can be written as:

$$\tilde{Q}'^*(\mu_{RS}) = \arg\max_Q [\Pi'_S + \mu_{RS} m^*(\mu_{RS})] = \arg\max_Q \Pi'_S$$

where $\Pi'_S = \underset{\varepsilon}{E}[(p_S - c)\min(Q, \mu'_S + \varepsilon) - c_k Q]$
s.t. $Q \geq \mu_{RS}$.

Corollary 2 For a given μ_{RS}, we have,

$$\tilde{Q}'^*(\mu_{RS}) = \max\left\{\mu_{RS}, E[T'\mu_{RS} + (1 - T')\mu'_S] + G'^{-1}\left(\frac{p_S - c - c_k}{p_S - c}\right)\right\}$$

Corollary 2 presents the supplier's reaction from the retailer's belief for a given value of μ_{RS}. Let $\tilde{\Pi}_R = \underset{\varepsilon}{E}[(p_R - p_S)\min(\tilde{Q}'^*(\mu_{RS}), \mu_R + \varepsilon)]$, the retailer's expected profit can be written as $\tilde{\Pi}_R - \mu_{RS} m(\mu_{RS})$. Therefore, the retailer's optimal ordered quantity $\tilde{\mu}^*_{RS}$ can be calculated by Eq. (11):

$$\tilde{\mu}^*_{RS} = \arg\max_{\mu_{RS}}[\tilde{\Pi}_R - \mu_{RS} m(\mu_{RS})] \tag{11}$$

where $\tilde{\Pi}_R = \underset{\varepsilon}{E}[(p_R - p_S)\min(\tilde{Q}'^*(\mu_{RS}), \mu_R + \varepsilon)]$.

(2) The supplier's decision.
After receiving the retailer's order $\tilde{\mu}^*_{RS}$, the supplier updates his forecasted demand to be $\mu(\tilde{\mu}^*_{RS})$ according to Eq. (2). The supplier's expected profit becomes $\tilde{\Pi}_S + m^* \tilde{\mu}^*_{RS}$, and his optimal capacity $\tilde{Q}^*$ can be written as:

$$\tilde{Q}^* = \arg\max_{Q}(\tilde{\Pi}_S + m^* \tilde{\mu}^*_{RS}) = \arg\max_{Q} \tilde{\Pi}_S$$

where $\tilde{\Pi}_S = \underset{\varepsilon}{E}[(p_S - c)\min(Q, \mu(\tilde{\mu}^*_{RS}) + \varepsilon) - c_k Q]$
s.t. $Q \geq \tilde{\mu}^*_{RS}$.

Corollary 3. For a given $\tilde{\mu}_{RS}^*$, we obtain

$$\tilde{Q}^*(\tilde{\mu}_{RS}^*) = \max\{\tilde{\mu}_{RS}^*, E[\tilde{\mu}_{RS}^* T(\tilde{\mu}_{RS}^*) + \mu_S(\tilde{\mu}_{RS}^*) - T(\tilde{\mu}_{RS}^*)\mu_S(\tilde{\mu}_{RS}^*)] + G^{-1}(\tfrac{p_S - c - c_k}{p_S - c})\}$$

Equation (11) and Corollary 3 present the engineering chain decisions when the soft-order is bounded. According to Eqs. (10–11) and Corollary 3, the equilibrium solution of retailer's soft-order, shared cost, and the supplier's capacity in coordination pattern are summarized as follows:

$$\begin{cases} \{\mu_{RS}^*, Q^*, \varphi\} & \text{if} \quad \tilde{\Pi}_S^* - \Pi_S^* > \Pi_R^* - \tilde{\Pi}_R^* \\ \{\tilde{\mu}_{RS}^*, \tilde{Q}^*(\tilde{\mu}_{RS}^*), m^*(\tilde{\mu}_{RS}^*)\} & \text{if} \quad \tilde{\Pi}_S^* - \Pi_S^* \leq \Pi_R^* - \tilde{\Pi}_R^* \end{cases} \tag{12}$$

Remarks: The retailer and supplier's decisions under the two-stage coordination process are calculated in Sect. 4.2. According to the retailer and supplier's binding conditions, it is obvious that both partners' expected profits in the proposed contract are no less than those in a decentralized pattern. Therefore, the two-stage coordination process is acceptable by both the retailer and the supplier.

The retailer makes decisions based on her belief of μ_S' since the supplier's forecasted demand μ_S is unknown to her. Simultaneously, the supplier makes decisions based on his evaluated market demand μ, while the real market demand is μ_R. Thus, the retailer and supplier do not know their real expected profits when they make their individual decisions. Since T, μ_S', and ε are random variables, an analytical study is prohibitively complicated. Therefore, we provide an experimental study to observe the roles that trust plays in decision making and examine the performance of our coordination contact.

2.4 Experimental Study

To find the effects of the trust on engineering chain decisions and the performance of the proposed trust-embedded cost-sharing coordination, we set up several scenarios in our experiment. The conclusions are as follows:

1. Evaluating Trust
   Observation 1. Many positive experiences are needed to gain trust, but a few negative experiences will lead to a big loss of trust.
   Observation 2. Trust-embedded coordination works since both the retailer and supplier value trust differently.
   Observation 3. The supplier's negotiation power does not necessarily mean profitability, i.e., a supplier's strong negotiation power might lead to a coordination failure.

The proposed trust-embedded coordination performs more efficiently when market demand volume is large. Since demand information distortion and demand volume fluctuations over time are common in industries, the proposed trust-embedded coordination is potentially helpful in practice.

3 ATP-Based Flexible Order Allocation Optimization in ATO Engineering Chain

3.1 Introduction

As a tool for enhancing the responsiveness of order promising and the reliability of order fulfillment, the available-to-promise (ATP) has increasingly attracted the attentions of the engineering chain managers and researchers. It directly links available resources, including both materials and production capacity, which affect the overall performance of an engineering chain.

In the traditional order fulfillment processes, the manufacturer makes order fulfillment plans after the arrival of new orders without consideration of resources availability. It might result in a high risk for the order fulfillments. In order to reduce the risk that some orders may not be fulfilled, companies have to keep large amounts of inventory. The rejections of some strategic customer orders or the high-profit orders may cause unbalanced usages of resources and ruin long-term interests. Therefore, it is important to ensure a high customer service level and maximize the profits by optimally allocating the key components and limited production capacity to the strategic customers or the high-profit orders.

From the current literatures (Xiong et al. 2003; Chen and Huang 2006; Yu-tao et al. 2008; Meyr 2009; Gao et al. 2012; Yang and Fung 2014) on available-to-promise and order fulfillment in engineering chain areas, although there have been some achievements in the research of ATP allocation and order fulfillment, the current literatures mainly focus on a single factory in MTS (make-to-stock) engineering chain. Very few papers consider the ATO and MTO engineering chain operation environment for a manufacturer who has multiple factories located in different areas. The study is also very limited on combining the ATP allocation model, which considers the customer priorities, and order fulfillment model, which is based on either batch orders or real-time orders. Based on the pre-allocation ATP of production capacity and components engineering capacity, we study an order fulfillment problem for a manufacturer with multiple factories in the ATO engineering chain. The order fulfillment models are established based on two kinds of order fulfillment mechanisms, i.e., the batch order fulfillment mechanism and the real-time order fulfillment mechanism. In the batch order fulfillment model, we propose a hybrid policy combining re-delivery and product substitution. In addition, the resource ATP (production capacity ATP and components engineering capacity ATP) searching rules are developed when the resource ATP is needed in the real-time order fulfillment model.

We focus on the optimization of the order fulfillment processes for a manufacturer with multiple production sites under the assemble-to-order (ATO) environment. Based on the constraints between the resources (production capacity and components) and customer demand priority level, a pre-allocated ATP model is established for the ATO engineering chain. Then, a batch fulfillment model (based on periodic operation) and a real-time fulfillment model (based on real-time operation) are presented by using the pre-allocated ATP results obtained from the pre-allocated ATP model.

3.2 ATP Pre-allocation Model

We consider a manufacturer, which has multiple factories and produces multiple products that are sold in multiple areas at different prices. Under the ATO engineering chain environment, the end products can be finished by simple assembling operations. And all factories can produce all kinds of products. However, due to the different cost on production equipment, labor, and other factors, the production cost and efficiency for the same product are different among factories.

In order to better achieve the manufacturer's development strategy and more profits, we classify the forecasting demand of next planned period into two levels according to historical sales data. An ATP pre-allocation model is built to allocate the component ATP and production capacity ATP for each demand level. Since that production is carried out only after orders arrive in the ATO engineering chain. To simplify the model, we assume that there is no initial stock of components and end products in the manufacturer at the beginning. Without loss of generality, we assume that the production lead time is assumed to be one period and the production preparation time and cost are negligible. And the components can be supplied at the beginning of each period.

The notations used in the ATP pre-allocation model are listed in Table 1.

In order to handle the difference between forecasting demand and actual demand in next period, we express the reserving rates of production capacity and components as cp and mp, respectively. The value of cp depends on the forecasting accuracy of demand quantity, while the

Table 1 Indices for ATP per-allocation model

Indices	Descriptions
f	Set of factories ($f \in F$)
m	Set of components ($m \in M$)
p	Set of products ($p \in P$)
t, τ	set of periods ($t \in T$, $\tau \in T$)
r	Set of demand levels ($r \in R$)
l	Set of selling areas ($l \in L$)

Table 2 Notations in the ATP pre-allocation model

Data	Descriptions
$Q_{\tau lrp}$	The total forecast demand quantity of product p coming from r demand level during period τ in selling area l
V_{lrp}	The unit profit of product p supplied to the r demand level in selling area l
$uscap_{fp}$	The consumed product capacity by factory f for making one unit of product p
$usmtrl_{pm}$	The consumed component m quantities for making one unit of product p
Cap_{tf}	The available product capacity in factory f during period t
$Matl_{tm}$	The greatest supplied quantities of component m during period t by vendor
$Cmsto_{fm}$	The inventory cost of component m in factory f each period per unit
$Cpsto_{fm}$	The inventory cost of product p in factory f each period per unit
$Ctran_{flp}$	The transportation cost of product p from factory f to selling area l per unit
Ie_{tfm}	The inventory of component m at the beginning of period t in factory f
It_{tfm}	The inventory of component m at the end of period t in factory f
Iep_{tfp}	The inventory of product p at the beginning of period t in factory f
Itp_{tfp}	The inventory of product p at the end of period t in factory f

value of mp depends on the forecasting accuracy level of demand for the variety of products. Then, some variables used in the ATP pre-allocation model are listed in Table 2.

Through the ATP pre-allocation model, the engineering plan, product capacity pre-allocation plan, component re-allocation plan, reserving product capacity plan, and reserving component plan can be obtained. The decision variables are shown in Table 3.

A mixed-integer programming model can be used to describe the re-allocated available-to-promise planning problem as follows:

$$\max\ Y^1 = \sum_{t=1}^{T}\sum_{f=1}^{F}\sum_{l=1}^{L}\sum_{r=1}^{R}\sum_{\tau=t}^{T}\sum_{p=1}^{P}(Qatp_{tflr\tau p}\cdot V_{lrp}) - \sum_{t=1}^{T}\sum_{f=1}^{F}\sum_{m=1}^{M} Cmsto_{fm}\cdot\left(\frac{Ie_{tfm}+It_{tfm}}{2}\right)$$
$$-\sum_{t=1}^{T}\sum_{f=1}^{F}\sum_{p=1}^{P} Cpsto_{fp}\cdot\left(\frac{Iep_{tfp}+Itp_{tfp}}{2}\right) - \sum_{t=1}^{T}\sum_{f=1}^{F}\sum_{l=1}^{L}\sum_{r=1}^{R}\sum_{\tau=t}^{T}\sum_{p=1}^{P}(Ctran_{flp}\cdot Qatp_{tflr\tau p}) \tag{13}$$

Table 3 Decision variables in ATP pre-allocation model

Decision variable	Descriptions
QP_{tfp}	The quantities of product p produced by factory f during period t
$QatP_{tflr\tau p}$	The quantities of product p supplied by factory f for the demand in period τ from r demand level in selling area l during period t
$SD_{f\tau p}$	The quantities of product p supplied by factory f for the demand in period τ
$pre\text{−}Matp_{tfr\tau m}$	The pre-allocated quantities of component m supplied by factory f for demand happened in period τ from r demand level during period t
$pre\text{−}Catp_{tfr\tau}$	The pre-allocated quantities of production capacity supplied by factory f for the demand in period τ from r demand level during period t
$Pmatl_{tfm}$	The quantities of component m received by factory f at the beginning of period t
mf_{tfm}	The real reserved quantities of component m in factory f during period t
cf_{tf}	The real reserved quantities of production capacity in factory f during period t

The first term is the profit derived from the anticipated sales based on the forecasting demand. The second term represents the inventory holding cost incurred for unused components when the component usage is less than the available supply during each period. The third term shows the inventory holding cost incurred for unsold products during each period. The last term represents the transportation cost from each factory to the demanding area.

The constraints on the pre-allocation resource problem for the each demand level are given in the following. Constraints (14)–(15) represent the demand restrictions. For the manufacturer, the production capacity and the available component supply are limited. Constraint (14) shows that the demand will be met as much as possible. The production quantities in every factory during each period are represented in constraint (15). In ATP pre-allocated plans, to avoid the delayed distribution, the product will not be allocated to the previous demand; this is represented by constraint (16).

$$\sum_{t=1}^{\tau}\sum_{f=1}^{F} Qatp_{tflr\,\tau p} \leq Q_{\tau lrp}, \quad \forall \tau, l, r, p \tag{14}$$

$$QP_{tfp} = \sum_{l=1}^{L}\sum_{r=1}^{R}\sum_{\tau=t}^{T} Qatp_{tflr\,\tau p}, \quad \forall f, p, t \tag{15}$$

$$\sum_{t=2}^{T}\sum_{f=1}^{F}\sum_{l=1}^{L}\sum_{r=1}^{R}\sum_{\tau=1}^{t-1} Qatp_{tflr\,\tau p} = 0, \tag{16}$$

Constraints (17)–(19) provide the production capacity restriction. Due to the finite production capacity, constraint (17) shows that the production quantities of each factory cannot exceed the available supply quantities during each period. Constraint (18) represents the real reserved quantities of production capacity. Constraint (19) gives the pre-allocated production capacity policy for each factory during different periods, which will be used by the real-time order fulfillment model.

$$\sum_{p=1}^{P} (QP_{tfp} \cdot uscap_{fp}) \leq Cap_{tf} \cdot (1 - cp), \quad \forall t, f \tag{17}$$

$$cf_{tf} = Cap_{tf} - \sum_{p=1}^{P} (QP_{tfp} \cdot uscap_{fp}), \quad \forall t, f \tag{18}$$

$$\sum_{l=1}^{L}\sum_{p=1}^{P} (Qatp_{tflr\,\tau p} \cdot uscap_{fp}) = pre - Catp_{tfr\,\tau}, \quad \forall t, f, r, \tau \tag{19}$$

Constraints (20–23) represent the components' restrictions. Constraint (20) provides the procurement plan which can be used by the batch order fulfillment model and the real-time order fulfillment model. In constraint (20), the production component reserved rate *mp* is used to balance the forecasting inaccuracy on product variety or the demand fluctuations. Constraint (21) indicates that the procurement quantities for each type of components cannot exceed the available supply capacity during each period. Constraint (22) gives the real reserved quantities for each component type. The pre-allocated component ATP policy can be obtained by constraint (23).

$$\sum_{p=1}^{P}(QP_{tfp} \cdot usmtrl_{pm}) \leq (1 - mp) \cdot Pmatl_{tfm}, \; \forall t, f, m \tag{20}$$

$$\sum_{f=1}^{F} Pmatl_{tfm} \leq Mat_{tm}, \; \forall t, m \tag{21}$$

$$mf_{tfm} = Pmatl_{tfm} - \sum_{p=1}^{P}(QP_{tfp} \cdot usmtrl_{pm}), \; \forall t, f, m \tag{22}$$

$$\sum_{l=1}^{L}\sum_{p=1}^{P}(Qatp_{tflr\tau p} \cdot usmtrl_{pm}) = pre - Matp_{tfr\tau m}, \; \forall t, f, r, \tau, m \tag{23}$$

Constraints (24–28) give the inventory levels of each type of component at the beginning and end of each period.

$$Ie_{1fm} = Pmatl_{1fm}, \; \forall f, m \tag{24}$$

$$Ie_{tfm} = It_{(t-1)fm} + Pmatl_{tfm}, \; \forall t \geq 2, f, m \tag{25}$$

$$It_{tfm} = Ie_{tfm} - \sum_{p=1}^{P}(QP_{tfp} \cdot usmtrl_{pm}), \; \forall t, f, m \tag{26}$$

$$Iep_{1fp} = 0, \; \forall f, p \tag{27}$$

$$Iep_{tfp} = Itp_{(t-1)fp}, \; \forall t \geq 2, f, p \tag{28}$$

In addition, decision variables $Pmatl_{tfm}$, QP_{tfp}, and $Qatp_{tflr\tau p}$ are integers and greater than or equal to zero.

We assume that products have the same functions but with different performances. Some customers would accept the substitution products if they can obtain some compensation. When the resource ATP is in shortage, the manufacturer can choose re-delivery order fulfillment policy, product substitution order fulfillment policy, or high-demand-level-priority order fulfillment policy to better meet customers' demand. The re-delivery order fulfillment policy shows that the manufacturer is able to meet customers' orders by two deliveries with the condition that customers allow receiving the products by two deliveries for the order and the quantity of the first delivery must meet the customers' request. The product substitution order fulfillment policy indicates that the manufacturer can meet customers' demand by substituting products partly when specified products are not available. The high-demand-level-priority order fulfillment policy says that pre-allocated production component ATP and capacity ATP for the lower demand level can be switched to higher demand level. In terms of the responses to the customers' orders, there are two order fulfillment models: the batch order fulfillment model and the real-time order fulfillment model.

In the batch order fulfillment model, all orders gathered during the batch interval will be fulfilled, and fulfillment date and quantities for each order will be given. The batch order fulfillment model can be used to determine whether or not the order can be accepted by the manufacturer and whether or not we need to apply the re-delivery order fulfillment policy or the product substitution order fulfillment policy, and the detailed fulfilled plan.

To avoid the situation that the orders of the strategic customers or the high-profit orders are rejected due to limited resources, in the real-time order fulfillment model, the high-demand-level-priority order fulfillment policy is adopted. With this policy, the higher level demand can use the pre-allocated production component ATP and capacity ATP that are already assigned to the lower level demand when the resource is in shortage. This real-time order fulfillment model will search the production capacity ATP along the backward time dimension, the demand level dimension, and the selling areas dimension. And this model will search the production component ATP along both forward and backward time dimension, the demand level dimension, and the selling areas dimension when components are in shortage.

3.3 Experimental Study

The proposed models are verified and tested through an example of the MP4 electronics manufacturing industry. The manufacturer has two factories and produces four kinds of products that have the same functions with different performances in two selling areas. The performance sequence of the four products is, from low to high, P1, P2, P3, and P4. In this paper, according to the customers' importance, we divide the customers into two levels and each level has a different priority.

To summarize, we can conclude the batch mechanism is better than the real-time mechanism on the order fulfillment. However, in order to response quickly, more and more manufacturers operate based on real-time order fulfillment mechanism. In addition, before orders are fulfilled, it is helpful for the manufacturer to optimize the pre-allocated resource ATP plan. And, the manufacturer can reserve some resources according to the inaccuracy of demand forecasting.

4 Conclusions

To remain competitive, firms must reduce operating costs while continuously improving customer service by coordinating and optimizing the overall engineering chain performances in the new business environment. Currently, engineering chain management is complicated and has its own special characters mainly reflected in two aspects. One is the random information from customer orders and the complex relations among engineering chain cooperators; the other is the outstanding relations of collaborative benefits and risks in this complicated environment. Therefore, we must probe the ways to respond to these characteristics and analyses, the coordination mechanisms, and optimization approaches in new business environment. We made a conjoint research of information sharing, operational research and behavior research to adaptive coordination mechanism. As for modeling and optimization, we focus on transferring orders to manufacturing and manufacturing customized orders under the ATO engineering chain environment, which is specific to fulfill the orders based on available production resources: materials and production capacity. Some conclusions are drawn as follows:

Since decision makers' feeling on trust is seldom studied in operations management area, investigating trust in engineering chain information sharing is meaningful. We first formulate a trust evaluation model based on psychology and statistics theories. The proposed model analytically explains that many positive experiences are needed to gain trust, but a few negative experiences will lead to a big loss of trust. Because of information asymmetry, both the engineering chain partners value trust differently and the engineering chain performs ineffectively. A coordination contract with a two-stage decision process is thereby proposed to coordinate the engineering chain. At the first stage, the supplier and retailer negotiate the cost-sharing rule. At the second stage, the retailer decides whether or not to bind her soft-orders and her optimal ordered quantity, while the supplier determines his optimal capacity. In order to maximize the supplier's expected profit, we find that there exists a crucial threshold of the supplier's negotiation power in negotiation, and strategically, making use of negotiation power is helpful to avoid business failures.

According to the pre-allocated production capacity and components, two order fulfillment models are formulated based on the batch processing and real-time mechanisms. In both order fulfillment models, an ATP searching method along the time dimension, the demand priority level dimension, the product dimension, or the selling area dimension is proposed when production capacity and components are in shortage. Several numerical examples are used to illustrate the proposed models. The experimental results show that the order fulfillment model with pre-allocation ATP is better than that without ATP pre-allocation. As a global optimization model, the batch mechanism is better than the real-time mechanism on the order fulfillments. When resources are in shortage, it is better to adopt the ATP searching method in the order fulfillment processes.

References

Cachon, G. P. (2003). Supply chain coordination with contracts. *Handbooks in Operations Research and Management Science, 11*, 227–339.

Cachon, G. P., & Lariviere, M. A. (2001). Contracting to assure supply: How to share demand forecasts in a supply chain. *Management Science, 47*(5), 629–646.

Cai, S., Goh, M., de Souza, R., & Li, G. (2013). Knowledge sharing in collaborative supply chain: Twin effects of trust and power. *International Journal of Production Research, 51*(7), 2060–2076.

Camarinha-Matos, L. M., & Macedo, P. (2010). A conceptual model of value systems in collaborative networks. *Journal of Intelligent Manufacturing, 21*(3), 287–299.

Chen, C.-T., & Huang, S.-F. (2006). Order-fulfillment ability analysis in the supply-chain system with fuzzy operation times. *International Journal of Production Economics, 101*(1), 185–193.

Chick, S. E., Huchzermeier, A., & Netessine, S. (2014). Europe's solution factories. *Harvard Business Review, 92*(4), 111–115.

Clemen, R. T., & Winkler, R. L. (1999). Combining probability distributions from experts in risk analysis. *Risk Analysis, 19*(2), 187–203.

Da Silveira, G., Borenstein, D., & Fogliatto, F. S. (2001). Mass customization: Literature review and research directions. *International Journal of Production Economics, 72*(1), 1–13.

Dolgui, A., & Proth, J.-M. (2010). *Supply chain engineering: Useful methods and techniques*. London: Springer Science & Business Media.

Duray, R., Ward, P. T., Milligan, G. W., & Berry, W. L. (2000). Approaches to mass customization: Configurations and empirical validation. *Journal of Operations Management, 18*(6), 605–625.

Ebrahim-Khanjari, N., Hopp, W., & Iravani, S. M. (2012). Trust and information sharing in supply chain. *Production and Operations Management, 21*(3), 444–464.

Fogliatto, F. S., da Silveira, G. J., & Borenstein, D. (2012). The mass customization decade: An updated review of the literature. *International Journal of Production Economics, 138*(1), 14–25.

Gao, L., Xu, S. H., & Ball, M. O. (2012). Managing an available-to-promise assembly system with dynamic short-term pseudo-order forecast. *Management Science, 58*(4), 770–790.

Ivanov, D., Pavlov, A., & Sokolov, B. (2014). Optimal distribution (re) planning in a centralized multi-stage supply network under conditions of the ripple effect and structure dynamics. *European Journal of Operational Research, 237*(2), 758–770.

Ivanov, D., Sokolov, B., & Pavlov, A. (2013). Dual problem formulation and its application to optimal redesign of an integrated production—distribution network with structure dynamics and ripple effect considerations. *International Journal of Production Research, 51*(18), 5386–5403.

Kosfeld, M., Heinrichs, M., Zak, P. J., Fischbacher, U., & Fehr, E. (2005). Oxytocin increases trust in humans. *Nature, 435*(7042), 673–676.

Laeequddin, M., Sahay, B., Sahay, V., & Abdul Waheed, K. (2012). Trust building in supply chain partners relationship: An integrated conceptual model. *Journal of Management Development, 31*(6), 550–564.

Meyr, H. (2009). *Customer segmentation, allocation planning and order promising in make-to-stock production. Supply chain planning* (pp. 1–28). Heidelberg: Springer.

Nike Inc. (1999). Annual Report.

Özer, Ö., & Wei, W. (2006). Strategic commitments for an optimal capacity decision under asymmetric forecast information. *Management Science, 52*(8), 1238–1257.

Özer, Ö., Zheng, Y., & Chen, K.-Y. (2011). Trust in forecast information sharing. *Management Science, 57*(6), 1111–1137.

Premkumar, G., Ramamurthy, K., & Nilakanta, S. (1994). Implementation of electronic data interchange: An innovation diffusion perspective. *Journal of Management Information Systems, 11*(2), 157–186.

Rousseau, D. M., Sitkin, S. B., Burt, R. S., & Camerer, C. (1998). Not so different after all: A cross-discipline view of trust. *Academy of Management Review, 23*(3), 393–404.

Sriram, M. (2005). Information asymmetry and trust: A framework for studying microfinance in India. *Vikalpa, 30*(4), 77.

Taylor, T. A., & Plambeck, E. L. (2007). Supply chain relationships and contracts: The impact of repeated interaction on capacity investment and procurement. *Management Science, 53*(10), 1577–1593.

Xiong, M., Tor, S. B., Khoo, L. P., & Chen, C.-H. (2003). A web-enhanced dynamic BOM-based available-to-promise system. *International Journal of Production Economics, 84*(2), 133–147.

Yang, W., & Fung, R. Y. (2014). An available-to-promise decision support system for a multi-site make-to-order production system. *International Journal of Production Research, 52*(14), 4253–4266.

Yu-tao, F., Sheng-chen, Y., & Yang, X. (2008). System for performing resource management and control and task scheduling in grid computing. In *International Symposium on Computer Science and Computational Technology, ISCSCT 2008.* IEEE.

Zhuge, H. (2011). Semantic linking through spaces for cyber-physical-socio intelligence: A methodology. *Artificial Intelligence, 175*(5), 988–1019.

Authors' Biography

Ming Dong Professor of Operations Management, Antai College of Economics & Management, Shanghai Jiao Tong University. His research and teaching interests include supply/value chain management, data-driven decision-making, operations management, and production optimisation.

Part III

Future Trends and Implications

11

Engineering for Sustainable Value

Miying Yang, Padmakshi Rana and Steve Evans

1 Introduction

In recent years, engineering companies are facing challenges from the depletion of resources, the rising price of material and energy, the environmental legislation, and the pressure from society (Evans et al. 2009). These challenges have been forcing engineering companies to develop new technologies and strategies to do business in a more sustainable way, in which less environmental and social negative impact is caused. Global production has affected economic and social development, as well as environment in a direct way. At a macro-level, industrial growth, globalisation, resource use (energy, water, and minerals), climate

M. Yang (✉) · P. Rana · S. Evans
Institute for Manufacturing, Cambridge University, Cambridge, UK
e-mail: my306@cam.ac.uk

P. Rana
e-mail: padmakshir@gmail.com

S. Evans
e-mail: se321@cam.ac.uk

Y. Zhang and M. Gregory (eds.), *Value Creation through Engineering Excellence*,
https://doi.org/10.1007/978-3-319-56336-7_11

change, household/consumer behaviour, and population growth amongst others have impact on high value engineering. At an industry (micro-) level, sustainability challenges refer to resource use and scarcity (human, physical, and financial), technology development, infrastructure design, workplaces (patterns of consumption), advertising and marketing to sell more stuff, and role of value and culture in shaping businesses and market (Michaelis 2003).

Engineering sustainability faces many challenges—from increasing scarcity of raw materials to reducing climate change impacts—that are broadly summed into the concept of sustainability. Without adequate responses to these challenges our highly engineered products will not be able to be produced and may not be able to be sold (Garetti and Taisch 2012). These challenges that shape the mainstream thinking on high value sustainable engineering require technical and business model changes. Companies are increasingly embracing approaches to sustainability such as eco-efficiency and clean technology. However, such innovations are not sufficient to address the pressing problems of unsustainability in engineering, in particular manufacturing. More fundamental changes in the way companies conceive and operate businesses are required. As such, a radical shift is required, where industry is considered pivotal in pursuing engineering for sustainable value.

The literature suggests (Lebel and Lorek 2008; Tukker and Tischner 2006) engineering for sustainable value requires systems approach that deals with systemic interdependencies and complexities and life cycle thinking, which takes into account products and processes view. Both approaches supplement each other and are considered necessary to address consumption and production patterns at different levels. Authors such as Evans et al. (2009) and Krantz (2010) suggest 'sustainability as an innovation platform' for a fundamental shift towards a sustainable economy with significant changes in people's lifestyle and mindset/behaviour, redesigning business models and value network 'to embrace a transformational sustainability that moves beyond incrementalism and eco-efficiencies'. Addressing these challenges both at macro-level and micro-level calls for significant changes to production and consumption, which requires participation and collaboration of companies, governments, non-government organisations, academia, and communities. The extended global value networks with multiple suppliers interacting with other networks and interrelationships between different industries through product use and disposal phase make collaborations between stakeholders across networks for innovating value integral to understand failed value exchanges. The World Economic Forum Report (2011) suggests that 'businesses are the builders of a sustainable consumption economy through their investments and innovation. The strategic use of life cycle thinking offers an opportunity to re-engineer business models and value chains'. Hence, such transition implies rethinking the business model to generate sustainable value (i.e. environmental, social, and economic value).

2 Literature Background

2.1 Sustainable Manufacturing

Industry, as estimated to be responsible for some 30% of CO_2 emissions on the planet, is a major consumer itself of primary resources and non-renewables and is the primary driver of end-user consumption of material goods (Evans et al. 2009). The huge impact on sustainability is also demonstrated by the relevance of energy consumption in manufacturing, primarily due to electrical energy and oil. Industry also develops and promotes demand for products that through their use cause significant additional CO_2 emissions and other forms of subsequent pollution and waste. Furthermore, the magnitude of the industrial sector, its global nature, use of natural resources for production, its role in technological innovation, its driving influence in most societies, and its primary position in a consumer-based culture make it central in impacting sustainability.

As Burke and Gaughran (2007) state, 'sustainability issues in manufacturing are growing exponentially. Initially referring to environmental considerations, sustainability now also encompasses social and economic responsibilities'. Garetti and Taisch (2012) define sustainable manufacturing as 'the ability to smartly use natural resources for manufacturing, by creating products and solutions that, thanks to new technology, regulatory measures and coherent social behaviours, are able to satisfy economic, environmental and social objectives'. Technological understanding is a part and not enough for a comprehensive view of sustainable manufacturing. Another perspective of sustainable manufacturing is related to consumer behaviour—people using smart products, services and in general for new technological capabilities in order to meet the sustainability challenges. Garetti and Taisch (2012) argued that sustainable consumption is part of sustainable manufacturing and emphasised that 'education is the prerequisite for consumer and people in general to correctly address the sustainability objectives through appropriate life-styles and the appropriate use of products and technology'.

Sustainable manufacturing has been the main focus of various researches supported by the European Commission, which has encouraged thinking of new perspectives in manufacturing linked to the sustainability challenges and, more recently, developing correspondent approaches, systems, and tools. Having concern to the new envisioned perspectives, it is worth mentioning research initiatives such as the IMS international project IMS 2020: supporting Global Research from IMS 2020 vision (IMS 2020 2009), which is in charge of preparing a roadmap for future manufacturing research and the Factory of the Future Strategic Multi-annual Roadmap (European Commission 2010), prepared by the Industrial Advisory Group for the Factories of the Future Public–Private Partnership. Furthermore, concern over social and environmental issues has resulted in a rising consumer pressure for responsible corporate behaviour has highlighted the need for responsible corporate behaviour to prove that complete focus on short-term financial results can lead the company towards jeopardy and total closure.

In summary, engineering for sustainable value requires changes to overall business processes and activities through collaboration amongst stakeholders in the value network. More specifically, a holistic solution focusing on redesigning business models and innovating value through the sustainability lens appears to be important. As Krantz (2010) proposes 'companies will need even bigger changes, including new business models, greater trust, and greater stakeholder engagement' based on a 'long-term vision' for pursuing sustainable consumption and production. Although environmental and social approaches are developed and implemented by companies, it is often through compliance with regulations or incremental environmental and social initiatives such as eco-efficiency, eco-innovation, and waste management and add-on corporate social responsibility activities in the community. These initiatives are helpful but incremental and limited in their ability to drive system-wide changes.

2.2 Business Model Innovation for Sustainability

Technological understanding is a part of engineering sustainability. It needs also suitable business models to achieve its commercial potentials (Chesbrough 2010). An increasing number of scholars and practitioner argue that technology and process innovation alone are no longer enough to create sustained competitive advantages, and the business model itself is key to unlocking the latent value potential of new technologies (Chesbrough and Rosenbloom 2002; Teece 2010). Recent research and practice show that business model innovation is a promising approach for improving sustainable manufacturing.

The concept of business model first appeared in the 1950s, but only became prevalent in the 1990s with the advent of the Internet (Teece 2010; Yip 2004). Business model in simple terms depicts 'how a firm does business' (Magretta 2002). All companies have some form of business model, even though they might not explicitly have considered or defined their model (Teece 2010). The concept of business model is closely linked to the concept of value in most business model literature. Majority of the existing literature defines business models in terms of value creation, capture, and delivery (Osterwalder and Pigneur 2010; Teece 2010). The literature (Chesbrough 2010; Zott and Amit 2010) suggests that business model innovation is a key to business success define business model innovation as 'a multi-stage process whereby organizations transform new ideas into improved business models in order to advance, compete and differentiate themselves successfully in their marketplace'. Björkdahl and Holmén (2013) regard business model innovation as *'a new integrated logic of how the firm creates value for its customers or users and how it captures value'*.

Lüdeke-freund (2010) defines a sustainable business model as '*a business model that creates competitive advantage through superior customer value and contributes to the sustainable development of the company and society*'. To develop sustainable business models, it is essential to consider the integration of social and environmental goals into a more holistic meaning of value (Schaltegger et al. 2012).

2.2.1 Product–Service Systems: A Pioneer of Sustainable Business Models

Product–service systems (PSS) is a set of business models that describe the selling off services rather than products alone. PSS is commonly classified into three types depending on the rate of service: product-oriented; use-oriented; and result-oriented PSS (Tukker 2004). Product-oriented PSS is that manufacturers sell products and provide added on services, for example, maintenance. Use-oriented PSS is that manufacturers sell utility of products instead of the physical products, such as products sharing, leasing and renting services. Result-oriented PSS is that manufacturers sell the result of products, such as selling printed documents instead of selling printers.

PSS business models are considered as promising ways to achieve sustainable production and consumption (Goedkoop et al. 1999; Maxwell and Van Der Vorst 2003; Mont 2002; Tukker 2015). The main reason is that engineering companies, in the context of PSS business models, have the incentive to prolong the lifetime of products and gain a long-term profit from service and end-of-life strategies, e.g. remanufacturing, reconditioning, repair, and recycling (Baines et al. 2007). It leads to a reduction of total material consumption throughout life cycle—dematerialisation, as well as a change of customers' consumption behaviour from buying products to buying services (Goedkoop et al. 1999). Therefore, PSS is regarded as a pioneer of sustainable business models with a potential to reorient both production and consumption towards a more sustainable direction (UNEP 2009).

All three PSS types have the potential of reducing environmental impact (Tukker and Tischner 2006). For example, the retaining of products ownership enables manufacturers to have the incentive to design for remanufacturing, recycle, reuse, and repair, which aligns with the purpose of sustainable design (UNEP 2009); the delivery of function or result increases the utilisation of products (Beuren et al. 2013). The use- and result-oriented PSS could deliver a higher potential to be dematerialised due to the retaining of ownership for manufacturer (Beuren et al. 2013) and thus are considered as the key to sustainable PSS (Roy 2000). Apart from environmental benefits, PSS also has the potential to be beneficial to society. For example, more jobs could be created from labour-intensive services (Beuren et al. 2013). However, it does not imply that PSS would inherently bring sustainable effects (Tukker and Tischner 2006).

The implementation of sustainable PSS business models is still challenging (Vezzoli et al. 2012). Sustainable PSS needs to be carefully designed at an early stage, since the design of PSS affects the material and energy consumption, cost, and customer behaviour through the entire life cycle (Ullman 2003). Various PSS development methods and tools have been proposed in the literature, for example Service Explorer (Sakao et al. 2009), Sustainable Product and/or Service Development (SPSD) (Maxwell and Van Der Vorst 2003), Methodology for Product–Service System Development (MEPSS) (Van Halen et al. 2005), and Solution-Oriented Partnership (Manzini et al. 2004). The existing methods and tools show that the development of sustainable PSS is still at an early stage. Few of the existing tools fully consider the social and environmental aspects of sustainability (Vasantha et al. 2012), which emphasises the need for methods and tools to support sustainable PSS development.

2.2.2 Cases of Sustainable Business Models

Two case studies in the manufacturing engineering sector are briefly presented here as examples of innovative business models that consider sustainability and generate sustainable value, despite being in a sector where continuous technological innovations and obsolescence tend to be key drivers for growth.

Riversimple

The car company is at an early start-up phase and was conceived to provide a personal and environmentally sustainable mobility solution encompassing technology solution and full service provision, adopting a total systems perspective. The company is based on a sale of service business model (PSS solution), which is about moving from resource consumption to resource efficiency. Current sales-based model rewards selling more and hence rewards the company directly for resource use; by shifting to a sale of service model the company retains ownership and responsibility of the vehicle and its operating costs for the product life and so is incentivised to design and build for durability, longevity, and efficiency in use, and end-of-life solutions. The company has an innovative governance model, where the company's stakeholder board elects the board of directors and executives. The stewards' board oversees the board of directors, and the custodian body represents the owners in limited partnership structure. This model is considered to assist in enhancing interactions and collaboration between stakeholders, to deliver sustainable value, by ensuring that financial interests are balanced with the interests of the other stakeholders.

It was observed that business modelling for sustainability in the company is ad hoc and driven by a visionary leadership. The breakthrough in the automotive industry, according to the founder, will come in the way a car is put together with the business model and delivery system (systems integration). It can be very powerful, particularly where there is a disruptive technology. The founder believes that for innovation in the automotive sector, the barriers are not really technological, but business and politics. Furthermore, the innovation is not in the individual component, but comes out of the synergy between the elements of the car (carbon fibre, fuel cells, ultra-capacitors, electric motors). However, with respect to the PSS solution there are significant questions around consumer adoption and ownership and how this might hinder the business model. The role of fashion and status and financial investment needs further understanding as these may represent significant barriers.

Elcon

Elcon is specialised to develop, market and produce uninterrupted AC and DC power systems, customised DC power supplies, DC/DC converters, custom tailored electronics and wireless solutions, while importing components for green power systems. The company's main focus is on solutions for energy and industrial plants. The battery backup systems are necessary to guarantee the 24/7 operation of critical devices also at any failure situations of the electrical mains network. Battery backed up DC power supply system solutions are being used in many power plants and stations, substations, and many other locations including, for example, an uninterrupted power supply of process automation. Elcon were interested in exploring and developing a service model for their business—new revenue streams through lease and reuse, and configuring the new value proposition with the potential new business model.

The development of business models for sustainability is a temporal process made up of a series of incremental activities building progressively towards a more completely integrated solution. It requires a long-term vision and focus on redesigning business models for value propositions that deliver sustainability. Companies adopt very different approaches to sustainability, but the common theme is that there is a business case for pursuing sustainability. Assisting companies in understanding the true scope of the impact of their activities on the broad range of stakeholders and identifying possible pathways to adaptation is only part of the challenge. A greater challenge is to persuade companies to do better when the business case is not so clear or when the payback period is unattractive.

In Elcon's case, 'the implementation of the new business model brought new challenges to the company's every-day operations. In order to respond to its value proposition, operational arrangements were made and new requirements for product development were identified. In this case operational changes were accomplished by networking with another manufacturing company. In this new setting the company's responsibilities are in operations related to sales, services, product and service design and development, while their partner is responsible for the manufacturing of the products.

2.3 Sustainable Value

At the core of the business model is the concept of generating value. The literature (Chesbrough and Rosenbloom 2002; Richardson 2008; Zott and Amit 2010) introduces the terminology of the 'value proposition' to describe the product/service offering that the company makes to its customers and other stakeholders for which it receives payment and aims to return a profit.

A holistic view of the value proposition requires active consideration of all stakeholders who are influenced directly or indirectly by activities of the firm (Rana et al. 2013). The key stakeholders discussed frequently in relation to sustainability include suppliers and partners, society, environment, suppliers, customers, investors and shareholders, governments, international organisations, non-government organisations (international and local) and the media. All business relationships include not only formal contractual activities, but also informal value exchanges of information and benefits. Greater visibility of all the value flows within a network potentially provides insights for innovation and improvement. Allee (2011) discusses the importance of tangible and intangible value flows in network. Understanding of intangible flows is important in understanding network relationships and identifying opportunities for further collaboration, including environmental and social aspects.

Sustainable value is defined as the well-being, improvement, continuity and preservation of the individual (human life), company, society and environment, in such a way that satisfies the needs of the present without compromising inter-generational equity. It is conceived as 'environmental' sustainability which covers sustainable use of natural resources, biodiversity conservation, recycling of waste and pollution, and provision of additional ecological services such as climate regulation, pollination, and enhancing soil fertility; 'social' sustainability is concerned with issues such as stakeholder participation, responsibility, labour standards, human rights, community relations, welfare, culture, poverty alleviation and equality; and 'economic' is concerned with traditional measures of financial profitability, risk management, and long-term economic viability or continuity of the company.

3 Conceptual Model of Engineering for Sustainable Value

In order to build a conceptual model of engineering for sustainable value, the authors further explored the existing literature and developed three key factors as below.

3.1 Factor 1: Life Cycle Value Creation

Life cycle thinking has been regarded as an essential concept for sustainable engineering in a holistic way. It seeks to identify possible solutions of improving goods and services by reducing resource use and environmental impacts throughout the entire product life cycle (European Commission 2011). The product life cycle can be divided into beginning of life (BOL), middle of life (MOL), and end of life (EOL) (Jun et al. 2007), as shown in Fig. 1. BOL is when the product is designed and manufactured; MOL is when the product is distributed and used; EOL is when the used product is reprocessed (e.g. recycled, reused, remanufactured) and disposed. Traditional manufacturers usually focus on the value creation in BOL since selling products is their main source of profit. The nature of sustainable business models extends the business relationship between manufacturer and customer from BOL to MOL and EOL, and thus brings more opportunities of value creation in MOL and EOL. For example, PSS business models could enable a long-term

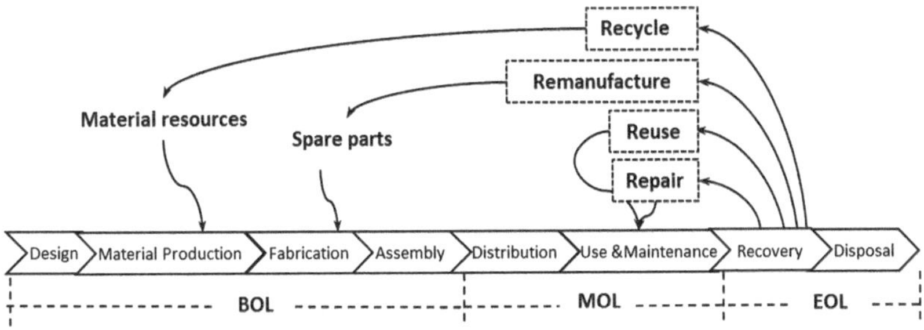

Fig. 1 Life cycle value creation, developed from (Jun et al. 2007)

profit for manufacturers from service, allow their access to data in use, and could achieve an improved utilisation of product. This motivates manufacturers to identify the opportunities of sustainable value creation in MOL and EOL (Toossi 2011).

3.2 Factor 2: Sustainable Value Analysis

Value refers to a broad set of benefits derived by a stakeholder from an exchange, which, in the context of sustainability, does not only include monetary profit, but also include social and environmental aspects (Rana et al. 2013). Figure 2 shows the three dimensions of sustainability and their interactions. Many researchers suggested that a sustainable manufacturing needs to take all three dimensions of sustainability into consideration (Maussang et al. 2009; Morelli 2002; Sakao et al. 2009). Sustainable value should cover all three dimensions, and sustainable value creation is proposed as a promising way of integrating sustainability into business modelling (Rana et al. 2013). Therefore, the concept of sustainable value could be integrated into engineering in order to also consider environmental and social aspects of benefits.

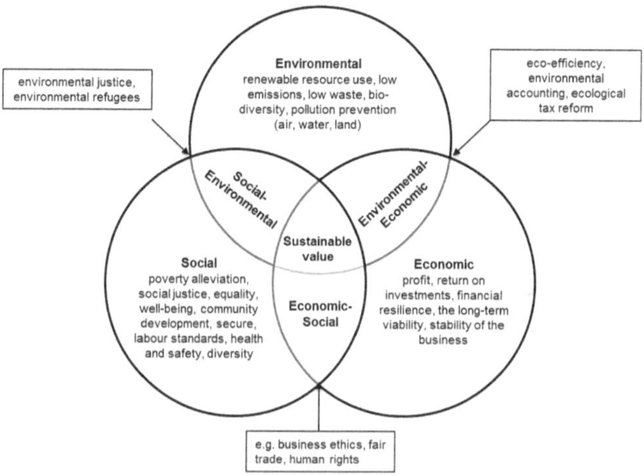

Fig. 2 Sustainable value (Yang 2015)

3.3 Factor 3: Comprehensive Forms of Value Analysis

This factor comes from the 'value mapping tool' for sustainable business modelling (Bocken et al. 2013). The value mapping tool has been successfully used in companies from various sectors, aiming to assist companies in the analysis and design of sustainable business model. This tool proposed the concepts of value destroyed and value missed to present the negative aspect of current business model. The rationale of this tool is that by analysing various forms of value companies can identify value creation opportunities through analysing value exchanges from the perspective of multiple stakeholders across the industrial network (Bocken et al. 2013; Short et al. 2012).

Based on this rationale, Yang (2016) further proposes value uncaptured, as a new perspective for sustainable value analysis. Value uncaptured is defined as the potential value which could be captured but has not been captured yet. Four forms of value uncaptured, i.e. value surplus, value absence, value destroyed and values missed and an approach of analysis of multiple forms of value were proposed as shown in Fig. 3 (Yang 2015; Yang et al. 2013).

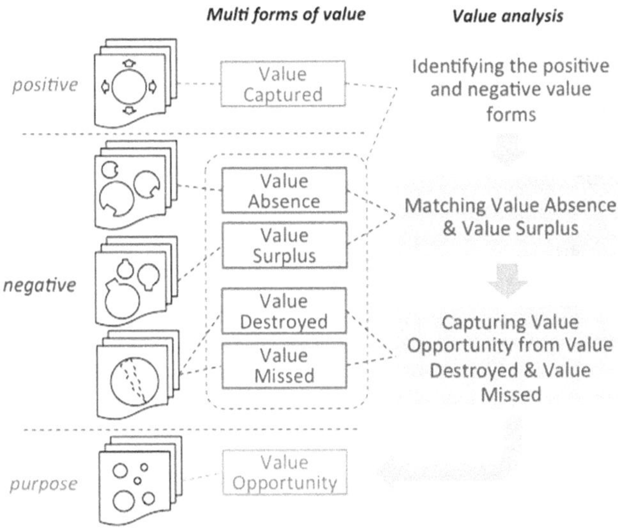

Fig. 3 Comprehensive forms of value analysis (Yang 2015; Yang et al. 2013)

Value uncaptured exists in almost all companies. Some uncaptured value is visible, e.g. waste streams in production, co-products, under-utilised resources, and reusable components of broken products; some is invisible, e.g. over capacity of labour, insufficient use of expertise and knowledge. Reducing any kind of the uncaptured value would create sustainable value. Figure 4 introduces the different forms of value uncaptured and explains how value uncaptured can trigger the identification of value opportunities.

The *Factor 1* and *Factor 2* are commonly presented in sustainability literature, while the *Factor 3* is newly proposed as a way of value innovation in our work. We believe that the analysis of multiple forms of value could be applied to support the development of sustainable engineering. Thus, we propose that the three factors could be combined to build a conceptual model of engineering for sustainable value as shown in Fig. 5.

This conceptual model aims to improve the understanding engineering for sustainable value by analysing comprehensive forms of value across the entire produce life cycle through the dimensions of economic, social, and environmental sustainability.

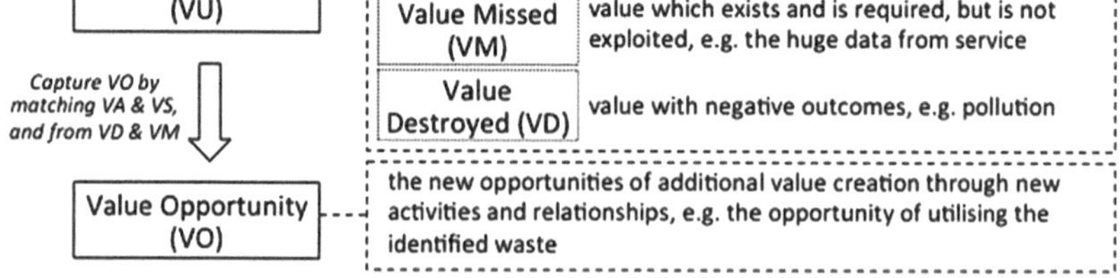

Fig. 4 Multiple forms of value uncaptured (Yang 2015)

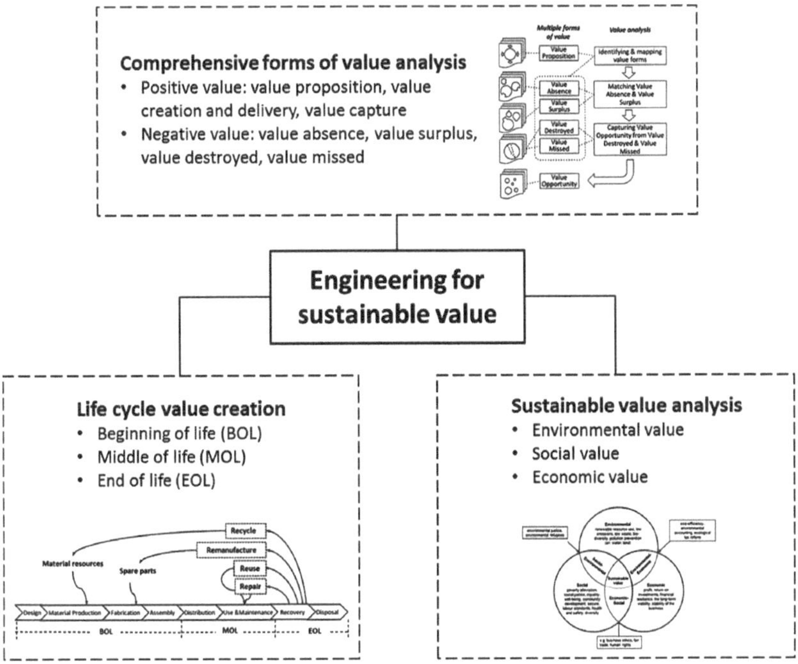

Fig. 5 Conceptual model of engineering for sustainable value

4 Sustainable Value Analysis Tool: A Tool for Engineering for Sustainable Value

Based on the conceptual model, SVAT is built to help manufacturers analyse multiple forms of value (including value uncaptured) across the entire product life cycle, and then to identify new opportunities for sustainable value creation. Identifying the value uncaptured and creating value from it is not always easy. SVAT is proposed to support this process, providing companies with a scheme to systematically look for each form of value uncaptured (i.e. value surplus, value absence, value destroyed and value missed) at the beginning, middle and end of the product life cycle, and with a method to turn the identified value uncaptured into value opportunities (Yang 2013).

Seven interviews and workshops with eleven managers/directors from five engineering companies were conducted to validate the conceptual model. Several common key feedbacks emerged:

- The life cycle thinking provides an extended view to look at the value creation opportunities at MOL and EOL, which is currently missed in most companies.
- The concept of value destroyed—negative value outcome—is clear. This concept could help identify the negative impacts to the environment and society. However, the concept of value missed needs further clearness—value currently squandered, wasted, or inadequately captured by current model. Besides, selling service is intangible, flexible, and unpredictable, and therefore requires a broader analysis on more value forms to identify the hidden value opportunities. The value surplus (e.g. waste) and value absence (e.g. need) proposed by the interviewers were regarded as helpful value forms.
- The conceptual model provides the interviewees a systems way of thinking about the economic, social, and environmental dimensions of a particular solution in the whole life cycle. A common interest to use a practical tool based on this conceptual model was raised from all of the five companies.

4.1 Development of Sustainable Value Analysis Tool

Based on the conceptual model and empirical validation in industries, SVAT is designed to help engineering companies identify opportunities to create sustainable value by analysing the captured and uncaptured value throughout the entire life cycle of products. The rationale of SVAT is to discover value opportunities by identifying and analysing value uncaptured (see Fig. 6).

It is not easy to identify value uncaptured in practice, so different forms of value uncaptured, i.e. value absence, value surplus, value destroyed, and value missed, is used to inspire the identification of value uncaptured. It is also difficult to discover value opportunities from value

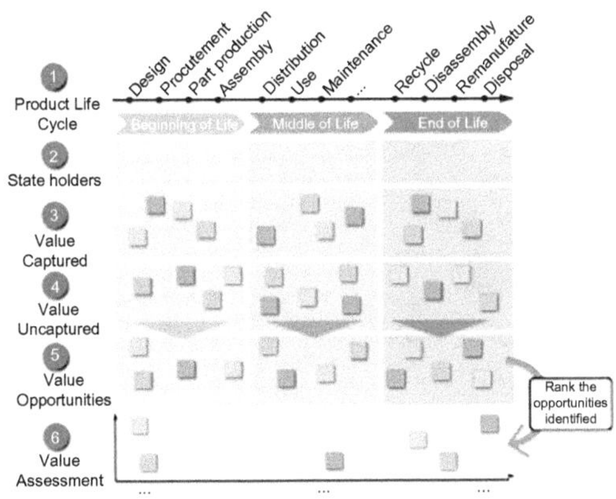

Fig. 6 Rationale of sustainable value analysis tool (Yang 2015)

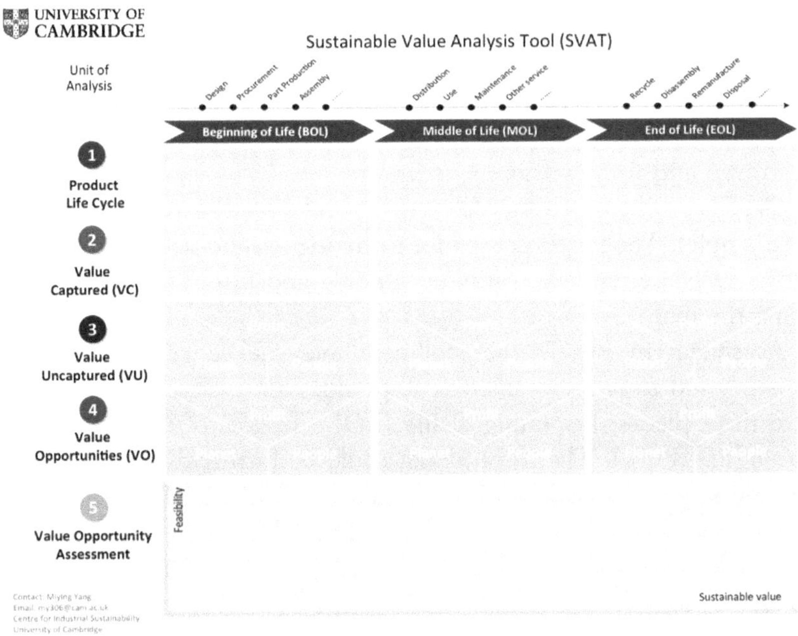

Fig. 7 Poster of sustainable value analysis tool (Yang 2015)

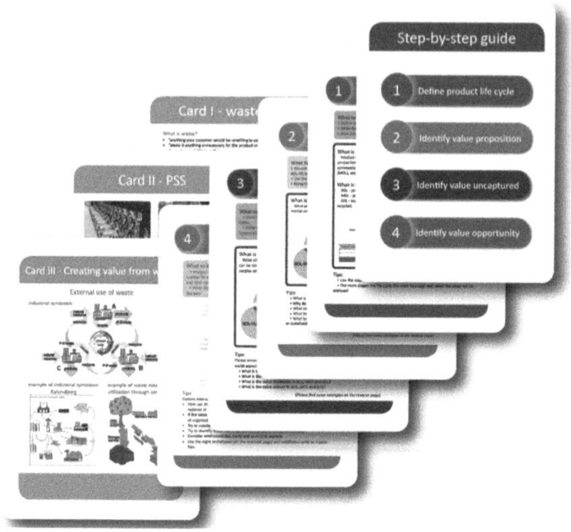

Fig. 8 Cards of sustainable value analysis tool (Yang 2015)

uncaptured, and different mechanisms are used to guide the process, e.g. by aligning value absence and value surplus, and reducing value destroyed and value missed.

SVAT consists of a poster (see Fig. 7) and a set of cards (see Fig. 8 for an example). The poster is used for gathering insights across the different life cycle phases and the cards for guiding and inspiring the process of using the tool.

As shown in Fig. 7, the tool combines the concepts of *life cycle thinking* and *value forms analysis*. A product life cycle could be divided into three phases: beginning of life (BOL), middle of life (MOL), and end of life (EOL). The three phases could be further divided into more specific stages. For example, MOL can be further divided into distribution, use, maintenance, and service. The value forms consist of value captured, value uncaptured, and value opportunities. Value uncaptured could be considered from the perspectives of value destroyed, value missed, value surplus, and value absence.

SVAT consists of five steps. For each step there is a card providing step-by-step guidance including background knowledge, tasks and tips on the front, and some inspirational examples on the back.

4.2 Use of Sustainable Value Analysis Tool

This SVAT can be used in facilitated workshops with managers, designers, or engineers in companies. It can be a stand-alone tool without support from other tools and can also be complementary to engineering tools or management tools. For the former case, the tool can be used to support decision-makings by identifying the value forms in the current business. For the latter case, the tool can be used during the conceptual design of product and service to support the integration of sustainability into PSS design.

Before using the tool

The industrial participants were asked to describe the current business model in their company. The researcher described the concepts and rationales underlying SVAT, explained the purpose of the tool and how to use it.

The process of using the tool

Step 1 *Define product life cycle*

The participants were asked to subdivide their product life cycle into more specific stages, depending on the depth of analysis desired and the time available. The more stages the life cycle is divided into, the more thoroughly the value will be analysed, but the more time-consuming it will be.

Step 2 *Describe the value captured*

Prompting questions and examples are provided on the card for Step 2. Users are encouraged to explain the economic, social, or environmental value they have identified, write them on post-it notes, and put them on the poster. Any value that involves more than one dimension should be placed on the boundary between them. For example, using recycled and healthy materials in BOL is considered valuable in terms of both social and environmental sustainability. These data therefore need to be posted on the boundary between the environmental and social dimensions.

Step 3 *Identify the value uncaptured*

This step included identifying the *value uncaptured* at each phase of the life cycle (i.e. BOL-VU, MOL-VU, and EOL-VU). The identification of *value uncaptured* involved identifying VS, VA, VD, and VM.

The researcher played a key role in helping the participants to identify the invisible and hidden *value uncaptured*. Firstly, the use of the tool requires both facilitator and practitioners to have substantial knowledge and experience of the entire product life cycle. Secondly, due to the intangibility and flexibility of service provision, the identification of hidden *value uncaptured* in MOL and EOL is difficult because the practitioners might not be aware of this concept. For example, many companies were not aware of redundant services as part of value destroyed. Therefore, facilitated brainstorming sessions were required, with practical examples, to provide inspiration and guidance.

Some *value uncaptured* could cover more than one dimension of sustainability. For example, 'no recycling and remanufacturing strategies' represents *value uncaptured* both in the economic and the environmental dimensions at the EOL stage. The particular advantage of considering the overlapping areas of the three dimensions is that it can help companies to identify *value uncaptured* across two or three dimensions of sustainability

Step 4 *Identify value opportunities*

Once the previous steps had been completed, the captured or uncaptured value was added to the table. The life cycle stage with a substantial concentration of value uncaptured could then be identified. The interviewer then helped the interviewees to analyse the root causes for this value uncaptured and the ways in which value might be created, inspiring the participants to explore opportunities for value creation.

It should be noted that the techniques used at this stage of tool development were still notional and lacked strong empirical evidence. This suggested there was a need to identify new techniques by building up experience of using the tool.

This step is similar to Step 4 for SVAT v1 and v2. The main difference is that by this stage the researcher had gained more facilitation experience and collected more examples and techniques to stimulate the users to identify value opportunities.

Step 5 *Assess the value opportunities*

The identified value opportunities (potential solutions) can be assessed briefly by discussing their impact on economic, social, and environmental areas. The opportunities offering high sustainable value can be selected and further analysed from the perspectives of feasibility, effectiveness, and ease of implementation. This provides a means of ranking the value opportunities.

It should be noted that in the context of sustainability, value does not only mean economic benefit, but also includes social and environmental aspects (Rana et al. 2013). Therefore, each value form needs to be considered from the three dimensions and their intersections.

The feasibility, usability, and utility of SVAT has been tested in a number of companies. Now the tool has been well received in 35 engineering companies across various sectors and of various sizes. It helped them find opportunities to create value internally and to discover the potential of creating mutual value externally. The tool has also been used for other purposes such as research, consultancy, business education, and university education.

5 Case Studies

The tool was used for case studies in an Air Separation Unit manufacturing company in China. The main business of this company is selling air separation units, petrochemical equipment, and industrial gases. This company has a yearly capability of designing and manufacturing more than 50 sets of large and medium air separation units sold to more than 40 countries and regions of the world. The reason of choosing this company is that it has successfully transformed from a traditional product-dominant manufacturing company to a company selling various types of PSS business models.

5.1 Business Model of This Company

Four main PSS business models have been implemented in this company. According to the classification of PSS types proposed by Tukker (2004), there are mainly four PSS solutions in this company as below.

1. *Technical service:* Apart from just selling air separation equipment to customers, this company provides technical services as added package to their products, such as maintenance, repair, and installation. This is the product-oriented PSS, and also the most common business model in this company.
2. *Special leasing:* This company leases the air separation units to customers, and the contract usually lasts for 10 years. During the contractual years, this company provides technical service to customers. After a certain amount of years, the ownership is transferred to the customer and a new contract will be made. Leasing contracts are not common in this company. They are mainly tailored for customers without the fanatical ability to buy equipment or build projects.
3. *Engineering procurement construction (EPC):* A subsidiary company was built to especially run EPC projects on April 2009. The main strategy of EPC in this company is extending the business from only selling a gas generator to selling an entire functional air separation system that customers need. The system includes the engineering system design, the procurement and production of facilities (e.g. refrigerator, compressor, fittings, rectifying tower, heat exchanger), the engineering construction, the installation of equipment, and related service (management, maintenance, etc.). EPC has generated big profit to the company and been regarded as their business trend.
4. *Industrial gas projects:* This is a result-oriented PSS that the company selling 'industrial gas' rather than 'gas generator'. This company

started the industrial gas projects in 2003 and has developed its own business model—combining air separation unit manufacturing and industrial gas management. Until 2012, 25 sub-gas companies have been built in 17 cities in China, reaching 980 km^3, covering the various industrial sectors, and producing gases such as O_2, N_2, CO_2, H_2, rare gases (Ar, He), and special gases. The investment is above 6.5 billion RMB. There are four commercial activities of providing gases: bottled gas, liquid gas and cold air separation of liquid, gasification, and pipeline industrial gas supply for industrial parks.

5.2 Using SVAT to Identify Value Opportunities in the Company

The SVAT was used with the director and two engineering designers from the engineering design sector of this company.

- The life cycle of the main products was defined as: BOL—customised design, procurement, part manufacturing, assembly; MOL—distribution, installation, use, maintenance, repair, management; EOL—disposal as scrap metal.
- The main value proposition was mainly identified at BOL and MOL phases. For example, the advanced technology has improved the energy efficiency; the customised design at BOL provided a better fulfilment of customer satisfaction; the PSS solutions brought long-term economic value to the company, and saved cost for customers, etc.
- The value uncaptured mainly existed at MOL and EOL phases. The company has implemented lean production, and there was little value uncaptured identified at the BOL. However, it has not taken any EOL strategies due to the limited market demand and high cost of recycling and remanufacturing. Also, the participants' awareness and knowledge of EOL is limited, so the value analysis in EOL is challenging. So, the

Table 1 Examples of identifying value opportunities from value uncaptured and their assessment

Value uncaptured	Value opportunities	Sustainability			Internal/External	Potential partners
		Economic	Environmental	Social		
Over-staffing	Reduce staffs	+	0	–	Internal	NA
	More work for company	+	0	0	Internal	NA
	More work for customer	+/0/–	0	+	External	Customer
Emissions of customer companies	Provide emission reduction solution to customer (RPSS)	±	+	+	External	Customer
Maintainers working far away from home if the projects are located in other cities	Hire local employees	±	+	+	External/Internal	Customer
	More holiday for far staff	–	–	+	Internal	NA
Co-products, e.g. N_2, O_2, Ar, liquid O_2, liquid N_2	Ar can be used for welding, and bulbs gas; O_2, N_2, etc. could be complementarily used for companies who need it;	+/0	+	0/+	External	Bulb company, etc.
Gold gas: Kr, Xe, etc.	The gold gases can be used in military industry, lamination;	+/0	+	0/+	External	Lamination company, etc.
High-purity O_2 and N_2	High-purity O_2 and N_2 can be used in electronic industry;	+/0	+	0/+	External	Electronic company

(continued)

Table 1 (continued)

Value uncaptured	Value opportunities	Sustainability			Internal/External	Potential partners
		Economic	Environmental	Social		
The waste of mechanical energy of expansion engines	The expansion movement can produce electricity to directly drive the engine itself or other engines.	+	+	0	Internal	NA
The waste of heat produced by the compressor	The wasted heat can be used to produce electricity or drive the compressor or vaporise the liquid O_2.	+	+	0	Internal/External	Steam turbine company

use of the tool is mainly focused on the identification of value uncaptured in MOL phase. The first column of Table 1 illustrates the selected identified value uncaptured at MOL in this company based on a sample of the data collected.
- Each of the main identified values uncaptured was analysed, and the value opportunities (potential solutions) were identified as shown in the table. For example, the value opportunity for the co-products (e.g. N_2, O_2, Ar, liquid O_2, liquid N_2) is identified to be that: Ar can be used for welding and bulbs gas (externally); O_2 and N_2 could be complementarily used for companies who need it (externally). The difficulty of this implementing the value opportunity is to identify and collaborate the 'external' company that the company can work with.

The tool has helped the company to identify value uncaptured and turn it into value opportunities. For example, the waste of low-grade heat and water was identified to be a major value uncaptured in the MOL, and the value opportunity is that it can be used to produce electricity or drive the compressor, or vaporise the liquid O_2. It can both be realised by the company internally or externally. Another example of value uncaptured is the waste of heat produced by the compressor. This is low-grade heat, which is difficult to reuse. The opportunity was identified to use the wasted heat to produce electricity or drive the compressor or vaporise the liquid O_2, which could bring positive economic and environmental impact. The company further analysed and found that a local steam turbine company could be the partner to implement it. Using the tool provides the company a broader vision that value opportunities could be identified by analysing the positive and negative aspects of the current business model.

6 Conclusion

This chapter proposes a conceptual model of engineering sustainability, using engineering for sustainable value and SVAT for this purpose. The concept of value uncaptured provides a novel way to improve engineering sustainability in the context of high value engineering.

The research contributes to theory by proposing that the combination of the three key factors—life cycle value creation, sustainable value analysis, and comprehensive forms of value analysis—could support the integration of sustainability into engineering operations. Life cycle thinking provides a holistic picture of the product from a design concept to the disposal, which allows for a system approach to examine the value creation in each stage. Sustainable value emphasises the combined consideration of environmental, social, and economic aspects of sustainability, which allows for the integration of sustainability concerns into value analysis. The multiple forms of value include value proposition and value uncaptured (i.e. value surplus, value absence, value missed, value destroyed), which allows for a thorough and comprehensive analysis of the positive and negative aspects of current business throughout the life cycle.

The chapter contributes to engineering operations practice by presenting the SVAT that integrates the three factors. This tool is developed to support engineering companies in their decision-making process to embed sustainability into the development of product–service systems. The tool is built upon a multi-disciplinary literature analysis and qualitative data from semi-structured interviews and workshops in five companies. The feedback of using the tool further confirms the need for developing a simple, usable, and workable tool for supporting the decision-makings in high value engineering areas, and for integrating sustainability into the consideration of this process.

References

Allee, V. (2011). *Value networks and the true nature of collaboration, ValueNet Works and Verna Allee Associates*. ValueNet Works and Verna Allee Associates.

Baines, T. S., Lightfoot, H. W., Evans, S., Neely, A., Greenough, R., Peppard, J., et al. (2007). State-of-the-art in product–service systems. *Proceedings of the Institution of Mechanical Engineers Part B Journal of Engineering Manufacture, 221*, 1543–1552.

Beuren, F. H., Ferreira, M. G. G., & Miguel, P. A. C. (2013). Product-service systems: A literature review on integrated products and services. *Journal of Cleaner Production, 47,* 222–231.

Björkdahl, J., & Holmén, M. (2013). Editorial: Business model innovation—the challenges ahead. *International Journal of Product Development, 18*(3), 213–225.

Bocken, N. M. P., Short, S. W., Rana, P., & Evans, S. (2013). A value mapping tool for sustainable business modelling. *Corporate Governance, 13*(5), 482–497.

Burke, S., & Gaughran, W. F. (2007). Developing a framework for sustainability management in engineering SMEs. *Robotics and Computer-Integrated Manufacturing, 23,* 696–703.

Chesbrough, H. (2010). Business model innovation: Opportunities and barriers. *Long Range Planning, 43,* 354–363.

Chesbrough, H., & Rosenbloom, R. S. (2002). The role of the business model in capturing value from innovation: Evidence from Xerox Corporation's technology spin-off companies. *Industrial and Corporate Change, 11,* 529–555.

European Commission. (2011). *Life cycle thinking and assessment—Our thinking—Life cycle thinking.*

Evans, S., Bergendahl, M., Gregory, M., & Ryan, C. (2009). *Towards a sustainable industrial system, International Manufacturing Professors Symposium in Cambridge UK.* Institute for Manufacturing, University of Cambridge.

Garetti, M., & Taisch, M. (2012). Sustainable manufacturing: Trends and research challenges. *Production Planning & Control, 23*(2–3), 83–104.

Goedkoop, M. J., Van Halen, C. J. G., Te Riele, H. R. M., & Rommens, P. J. M. (1999). *Product service systems, ecological and economic basics, Ecological and Economic Basics.* The Hague: Ministry of Environment.

Jun, H., Kiritsis, D., & Xirouchakis, P. (2007). Research issues on closed-loop PLM. *Computers in Industry, 58*(8–9), 855–868.

Krantz, R. (2010). A new vision of sustainable consumption: The business challenge. *Journal of Industrial Ecology, 14*(1), 7–9.

Lebel, L., & Lorek, S. (2008). Enabling sustainable production-consumption systems. *Annual Review of Environment and Resources, 33*(1), 241–275.

Lüdeke-Freund, F. (2010). Towards a conceptual framework of business models for sustainability. *Environmental Management, 49,* 1–28.

Magretta, J. (2002). Why business models matter. *Harvard Business Review, 80*(5), 86–92.

Manzini, E., Collina, L., & Evans, S. (2004). *Solution oriented partnership—How to design industrialised sustainable solutions.*

Maussang, N., Zwolinski, P., & Brissaud, D. (2009). Product–service system design methodology: From the PSS architecture design to the products specifications. *Journal of Engineering Design, 20*(4), 349–366.

Maxwell, D., & Van Der Vorst, R. (2003). Developing sustainable products and services. *Journal of Cleaner Production, 11*(8), 883–895.

Michaelis, L. (2003). The role of business in sustainable consumption. *Journal of Cleaner Production, 11*(8), 915–921.

Mont, O. (2002). Clarifying the concept of product–service system. *Journal of Cleaner Production, 10*(3), 237–245.

Morelli, N. (2002). Designing product/service systems: A methodological exploration. *Design Issues, MIT Press, 18*(3), 3–17.

Osterwalder, A., & Pigneur, Y. (2010). *Business model generation: A handbook for visionaries, game changers, and challengers,* (Vol. *5*).

Rana, P., Short, S., & Evans, S. (2013). *D2.5 Lessons learned report, documenting the impact from use of the tools & methods and areas for improvement.*

Richardson, J. (2008). The business model: An integrative framework for strategy execution. *Strategic Change, 17,* 133–144.

Roy, R. (2000). Sustainable product-service systems. *Futures, Unknown, 32,* 289–299.

Sakao, T., Sandström, G. Ö., & Matzen, D. (2009a). Framing research for service orientation of manufacturers through PSS approaches. *Journal of Manufacturing Technology Management, 20*(5), 754–778.

Sakao, T., Shimomura, Y., Sundin, E., & Comstock, M. (2009b). Modeling design objects in CAD system for service/product engineering. *Computer-Aided Design, Elsevier Ltd, 41*(3), 197–213.

Schaltegger, S., Lüdeke-Freund, F., & Hansen, E. (2012). Business cases for sustainability and the role of business model innovation. *International Journal of Innovation and Sustainable Development, 6*(2), 95–119.

Short, S.W., Bocken, N.M.P., Rana, P., & Evans, S. (2012). Business model innovation for embedding sustainability: A practice-based approach introducing business model archetypes. In *The 10th Global Conference on Sustainable Manufacturing (GCSM): Towards Implementing Sustainable Manufacturing.* Istanbul, Turkey.

Teece, D. J. (2010). Business models, business strategy and innovation. *Long Range Planning, 43*(2–3), 172–194.

Toossi, A. (2011). *A value-centric decision making framework for maintenance services outsourcing*. Cranfield: Cranfield University.
Tukker, A. (2004). Eight types of product–service system: Eight ways to sustainability? Experiences from SusProNet. *Business Strategy and the Environment, 13*(4), 246–260.
Tukker, A. (2015). Product services for a resource-efficient and circular economy—A review. *Journal of Cleaner Production, 97,* 76–91.
Tukker, A., & Tischner, U. (2006). *New business for old Europe: Product-service development*. Shipley: Greenleaf Publishing Limited.
Ullman, D. G. (2003). The mechanical design process. *Mechanics of materials* (Vol. 3, p. 432). New York: McGraw-Hill.
UNEP. (2009). *Design for sustainability: A step-by-step approach.*
Van Halen, C., Vezzoli, C., & Wimmer, R. (2005). *Methodology for product service system innovation: How to develop clean clever and competitive strategies in companies* (Vol. 75). Assen: Koninklijke Van Gorcum.
Vasantha, G. V. A., Roy, R., Lelah, A., & Brissaud, D. (2012). A review of product–service systems design methodologies. *Journal of Engineering Design, 23*(9), 635–659.
Vezzoli, C., Ceschin, F., Diehl, J. C., & Kohtala, C. (2012). Why have sustainable product–service systems not been widely implemented? *Journal of Cleaner Production, Elsevier Ltd, 35,* 288–290.
Yang, M. (2013). *Sustainable value analysis tool.*
Yang, M. (2015). *Sustainable value analysis for product–service systems*. University of Cambridge.
Yang, M., Evans, S., Vladimirova, D., & Rana, P. (2016). Value uncaptured perspective for sustainable business model innovation. *Journal of Cleaner Production*. doi:10.1016/j.jclepro.2016.07.102.
Yang, M., Rana, P., & Evans, S. (2013). Product service system (PSS) life cycle value analysis for sustainability. In *The 6th International Conference on Design and Manufacture for Sustainable Development (ICDMSD).*
Yip, G. S. (2004). Using strategy to change your business model. *Business Strategy Review, 15*(2), 17–24.
Zott, C., & Amit, R. (2010). Business model design: An activity system perspective. *Long Range Planning, 43,* 216–226.

Authors' Biography

Miying Yang is Lecturer of Engineering Management at the University of Exeter. She holds a PhD from the Department of Engineering at the University of Cambridge. Her research is focused on sustainable business model innovation, sustainable value creation and servitisation. She is interested in transforming theories into practical tools that help industries solve real problems.

Padmakshi Rana is the co-founder of the Inceptery, a sustainability-led innovation consultancy. She worked as a writer for Eco-Business, an online sustainability publication. Prior to this, she was a research associate at the Centre for Industrial Sustainability, University of Cambridge. Her research interests and consultancy work focus on sustainable business models, value exchanges and networks, stakeholder collaboration, materiality and reporting for sustainability.

Steve Evans is professor of sustainable manufacturing and Director of the National EPSRC Centre for Industrial Sustainability, conducting research into how industry can bring ambitious environmental sustainability goals into practice. He was previously Engineering Systems Manager at Martin-Baker Engineering. He is a co-founder of multiple start-ups and has supported various governments in their transition towards a sustainable industrial system, including as Specialist Advisor to the House of Lords in the UK.

12

Product Life Cycle Design for Sustainable Value Creation

Jing Tao and Suiran Yu

1 Introduction

Value is the reliable performance of functions to meet customer needs at the lowest overall cost, and it can be calculated as (SAVE 2007):

$$\text{Value} = \text{Function/Cost}$$

where Function is what the product or service is supposed to do, while Cost is the expenditure needed to create it. These three characteristics, denominated as the "survival tripod" by Cooper and Slagmulder (1997), are related as a rule for the success of companies, which should balance this tripod in accordance with market requirements and the company strategy.

J. Tao (✉) · S. Yu
School of Mechanical Engineering, Shanghai Jiao Tong University, Shanghai, China
e-mail: j_tao@sjtu.edu.cn

S. Yu
e-mail: sryu@sjtu.edu.cn

Y. Zhang and M. Gregory (eds.), *Value Creation through Engineering Excellence*,
https://doi.org/10.1007/978-3-319-56336-7_12

However, the question arises as existing value concept and definitions sufficient when viewing values form broader perspectives of sustainability. Just to quote: Long-term thinking has to be instilled; old fashioned values have to be recovered so that a sustainable growth and survival becomes realistic again (Ostad Ahmad Ghorabi and Jerlich 2008). Sustainable value refers to a broad set of benefits derived by a stakeholder from an exchange, which, in the context of engineering sustainability, do not only include monetary profit, but also include social and environmental aspects (Adams 2006; Rana et al. 2013; Yang et al. 2014). Sustainable value should cover all three dimensions, and sustainable value creation is proposed as a promising way of integrating engineering sustainability into the life cycle systems. Therefore, the concept of sustainable value creation could be integrated into business in order to also consider environmental and social aspects of benefits. Figge and Hahn (2013) also proposed a sustainable value model that aims at the quantitative assessment of the value-creating use of environmental, economic, and social resources. The approach can be used to answer the financial-economic question of "where environmental and social resources should be allocated in order to achieve an optimal overall return". This value-oriented approach is necessarily complementary with burden-oriented approaches, and both need to be considered to arrive at an optimal allocation of resources (Merante et al. 2015). The equation for sustainable value calculation is as follows:

$$\mathrm{SV} = 1/R\sum_{r=1}^{R} x_{ir}\left[\left(\frac{y_i}{x_{ir}}\right) - \left(\frac{y^*}{x_r^*}\right)\right] \tag{1}$$

where SV refers to the sustainable value of the evaluated system, R stands for the number of resources considered in the evaluation, r stands for the individual resource (e.g. water, land, energy), y_i and y^* stand for the value added of the evaluated system and the benchmark, respectively, and, finally, x_{ir} and $x_r{}^*$ stand for the amount of resources used by the evaluated system and the benchmark, respectively. The SV indicates the extent to which a system contributes to make the resource use more sustainable. To achieve this, the efficiency use of the company's resource is compared against the efficiency use of the same resource at the benchmark level, such as a national economy, an industry sector, another company, or a performance target.

This chapter focuses on sustainable value creation in the context of high value engineering. First, the framework of sustainable value-driven life cycle design is presented for product and process engineering innovation for sustainability and provides the conceptual linkage with the value-creating activities of the firm such as design, production, supply chains, partnerships, and distribution channels. Then, approaches for the integration of ecological assessment (i.e. LCA) with computer-aided product development are proposed as a useful tool to support sustainability-oriented product and process engineering. The proposed approaches and tools are expected to help bring experts in fields of product and process engineering, industrial management, and ecological assessments to a common vision, and to accelerate development of more sustainable products, processes, and business strategies.

2 Product Life Cycle Design for Sustainable Value Creation

Sustainable product development helps to use a company's resources, usually in terms of materials and energy in the most economical way which implies the least environmental impact. This constitutes a benefit for both, the organization and their customers at the same time. Life cycle design (LCD) is considered to be a promising approach for reduction of environmental impacts and promotion of product performance throughout its life cycle. The term LCD refers to an integrated design of a product and its life cycle from material extraction to disposal, even recovery. A sustainable product life cycle is defined as a life cycle which can minimize the material and energy consumption, amount of waste, and environmental emissions from the viewpoint of the whole life cycle while fulfilling function, quality, cost, and profit requirements. Thus, the life cycle offers a framework for innovation for sustainability and provides the conceptual linkage with the activities of the firm such as design, production, supply chains, partnerships, and distribution channels. For supporting the life cycle design, various concepts and Design for X (DfX) methodologies have been proposed; examples include industrial ecology (Graede and Alleby 1995), life cycle planning (Ishii et al. 1997; Kato et al. 2000; Kato and Kimura 2004; Kobayashi 2006; Umeda 2001), life cycle costing (LCC), design for the environment (Ray and Guazzo 1993; Zhang et al. 2011), design for disassembly (Kroll and Hanft 1998; Noller 1992), design for reuse and remanufacturing (Du et al. 2013; Umeda et al. 2006; Zwolinski et al. 2006), design for recyclability (Ishii and Lee 1996; Rose et al. 1998), end-of-life design (Kara et al. 2005; Umeda et al. 2006; Xing et al. 2003; Ziout et al. 2014), and modular design considering life cycle issues (Yu et al. 2011).

This study focused on the methods and tools for sustainable values creation and promotion in the context of product and process engineering. Given the advantages of LCD, this study tied to integrate the sustainable value concepts and LCD theory for development of sustainable value-driven product life cycle design methodologies. In order to accomplish this goal, three research problems are to be addressed. First, identification and description of the various sustainable value opportunities throughout the product life cycle. In life cycle design, a designer should consider various aspects of sustainability, including resource and energy efficiency, wastes and emissions, economic benefits and social benefits. Thus, the comprehensive understanding of design-driving factors is the very first and important step of all the design processes. In this step, it is important to select and carefully formulate the sustainable design goal indicator system, because these indicators are important to assessment of the design results and conclusion of the design process. Second is the generation of sustainable life cycle solution by translation of various sustainable value goals into operative engineering characteristics of product and its LC process. The design team proposes a set of life cycle design solutions that can achieve the design goals, focusing on interrelations among the goals, the product, and the life cycle flow. Though many researchers identified the importance of strategies required to integrate product design and life cycle processes in life cycle engineering (Ishii 1995; Ometto et al. 2008; Tchertchian et al. 2010), they mainly focused on the end-of-life strategy to close the material loop and design methods for realizing the end-of-life strategy. Therefore, an integrated life cycle design methodology must help engineers to estimate the life cycle implication of a candidate design, identify cost and profit drivers, and facilitate "simultaneous" design of the product and enlistment of the manufacturing specifications, service logistics, and product retirement plan throughout the engineering value chain associated with the life cycle. Third, design evaluation methods and tools. Life cycle assessment (LCA) is a strong tool for evaluating material and energy consumption and emissions of a life cycle. However, it cannot evaluate balances of a life cycle in terms of material, energy, and money, especially when the life cycle has loops such as remanufacturing, reuse, and recycling. LCA itself could not provide the rate of material circulation by means of, for example, part reuse, as it depends on various factors such as failure rate of the parts, the market life of the product, and the efficiency of the collection system, and the relation between these factors is complicated and cannot be represented by a simple mathematical model. Therefore, more powerful modelling and analysis supporting tools are needed.

2.1 Framework for Sustainable Value-Driven Product Life Cycle Design

Shown in Fig. 1 is the framework for sustainable value-driven life cycle design proposed by this study. The concept of "domain" (Suh 2001) is introduced. Three design domains which are the sustainability goals (SGs) domain, the life cycle function requirements (LcFRs) domain, and the LCD solution domain are then constructed. Based on the definition of the three domain, LCD is then organized as the structured and strategic mapping of SGs to functional requirements of both product and it major life cycle processes, and then simultaneously into operational engineering characteristics of both product and its major life cycle process. A three-dimensional model of sustainable value goals for life cycle design is first proposed. Then, a QFD-based life cycle scheming approach is proposed for sustainable product life cycle strategy formulation. Also, Life Cycle Simulation is then employed for modelling

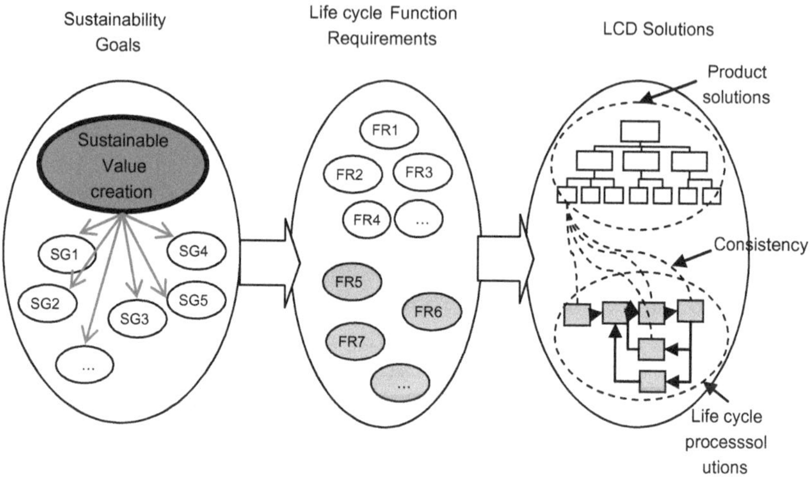

Fig. 1 Sustainable value-driven life cycle design framework

complicated closed-loop-type product life cycles and quantified evaluation of sustainable value potentials of different stages and stakeholders.

For readers' comprehension of the sustainable value-driven life cycle design framework, the design elements in each domain are explained as follows:

- Sustainability goals of life cycle design

It is proposed in the literature that value has multiple forms, including value destroyed, value missed, and value opportunity. Different forms of value can be converted into each other based on the mechanism between them (Bocken et al. 2013). However, the ultimate goal of value engineering is value creation. In order to map the sustainable value into the framework of life cycle engineering, the concept of sustainability goals for life cycle design is proposed in this paper. Sustainability goals (SGs) are the identified requirements on various potentially sustainable value-creating engineering activities throughout the engineering value chain associated with the product life cycle. Therefore, based on the definition of sustainable value, a three-dimensional model of SGs for life cycle design is proposed as shown in Fig. 2. First, SGs are identified

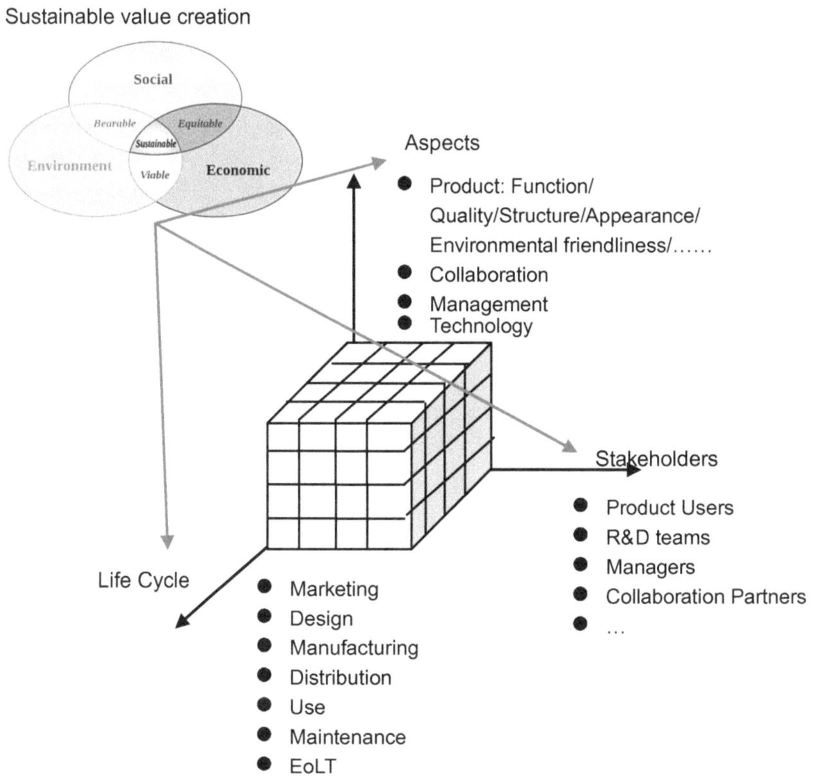

Fig. 2 Three-dimensional model of SGs

throughout the life cycle from R&D to end-of-life treatment (EoLT). It includes not only function and performance information required for product design and manufacturing, but also information of products status and users' feedbacks during product use, and requirements on product end-of-life treatments. Second, SGs are identified from perspectives of different stakeholders associated with the life cycle, including product users, collaboration partners, managers, R&D teams. Third, SGs can be categorized into several different aspects including product and process engineering, business and management, and therefore has the strong capability of requirement description.

SGs can be categorized into requirements on product itself including function and performance, quality, structure, appearance, material, cost, etc. Collaboration requirements include those on outsourcing, collaborative design, planning and business, etc. Technological requirements include manufacturability flexibility, procession, environmental-friendliness, etc. Management requirements include process optimization, security, law and regulation obedience, cost and profit, etc.

- Life cycle function requirements (LcFRs):

Here, function requirements include those on both product and its life cycle process. For instance, function requirements of mechanical products generally include those on energy, motion, control, material, and environmental impacts (see Fig. 3).

Compared with traditional product design, the most important and distinguishing feature of life cycle design is the design of the circulation of products, components, and materials around the life cycle so as to minimize resource demands and environmental emissions. In order to realize life cycle design objectives, life cycle stage of maintenance and end-of-life treatment should be carefully considered and planned. Therefore, function requirements on maintenance and end-of-life stages of product life cycle were illustrated (see Fig. 4).

- Sustainable LCD solutions:

LCD solutions refer to product and life cycle process design solutions. Product design solutions (ProdS) generally mean the basic structure of

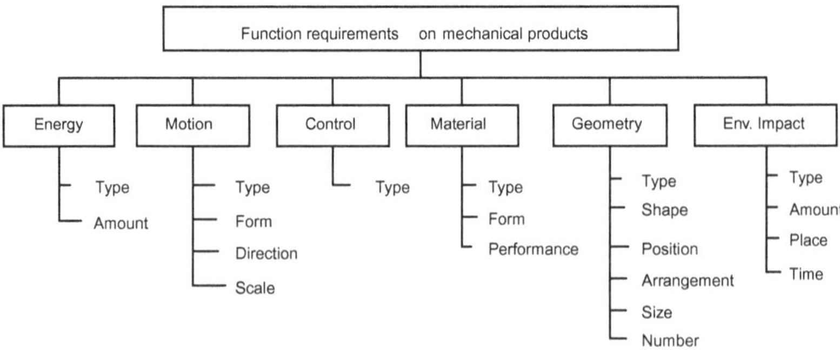

Fig. 3 Function requirements on mechanical products

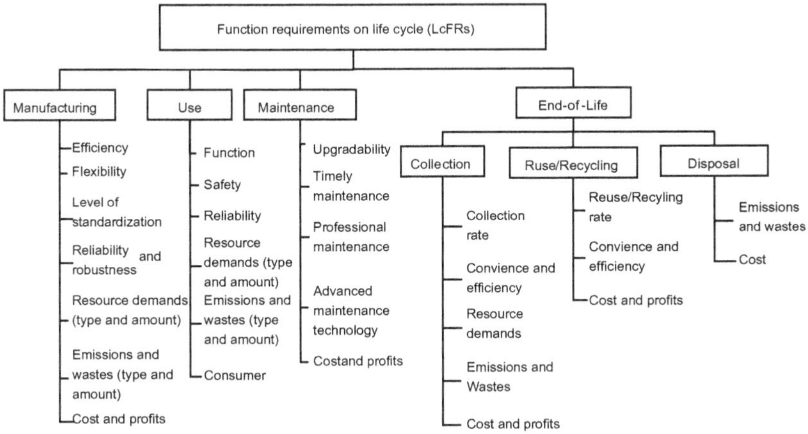

Fig. 4 Function requirements on life cycle

product such as major components and their key engineering characteristics such as material, weight, shape, etc. The product structure can be typically demonstrated by the hierarchical model as shown in Fig. 5.

The life cycle process solution here is defined as a combination of life cycle options (LOPs) such as sustainable manufacturing, maintenance, collection, recycle, and reuse of disposed products and components and describe the flows of products, components, and materials around the life

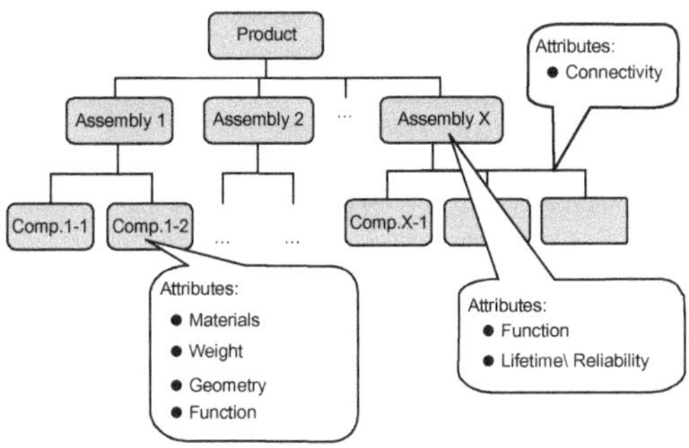

Fig. 5 Hierarchical product model

cycle. And each option is further described by certain attributes (see Table 1). Based on the plan, the life cycle as a network of life cycle unit processes can be developed.

Table 1 Life cycle options and attributes

LC options (LOPs)	Attributes
Sustainable manufacturing	Manufacturing technology choice Process chain Energy efficiency improvements (%) Use ratio of recycled material (%wt) Use ratio of reconditioned components (%) Monitoring and surveillance methods Pollutions prevention strategy Reduction of wastes and emissions (%) ……
Distribution and use	Purchase Leasing Use mode design Logistics service choice Pricing strategy ……
Maintenance	Maintenance strategy Upgrade strategy Guarantee time Estimated maintenance preference of end-users Level of maintenance Maintenance and preventive check cycle Training of maintenance stuff Specialized equipment and tools for maintenance ……
Collection of used products and components	Collection plans Logistics service choice Pricing strategy ……
Reuse/recycling	Disassembly plan Examining and inspection technologies Refurbishing/recycling technologies Reuse choice: installation reuse/maintenance reuse/global reuse Recycling choice: product manufacturing/global treatment of residues ……

2.2 Generation of Sustainable Life Cycle Design Solutions by a Modified QFD Approach

In order to realize the LCD process proposed in the last section, a modified QFD method is developed in this paper. Applications to describe environmental requirements in a QFD matrix have been proposed (Masui et al. 2001; Sakao 2007; Sakao et al. 2001; Zhang et al. 1999). However, the methods are still focused on the design of product itself. The proposal of the QFD method in this paper is inspired by the study of An et al. (2008) on integrated product–service roadmap. As shown in Fig. 6, the modified QFD creates two interlinked mappings. The first mapping starts with SGs (inputs) and then translates these requirements into LcFRs. The second mapping follows through these LcFRs and translates them into the product characteristics (generally speaking, the structure elements, i.e. module, part, component, of the product) and LC process options and attributes (outputs).

Compared with the traditional QFD, several modifications have been done for the proposed approach. Similar to the study of An et al., two

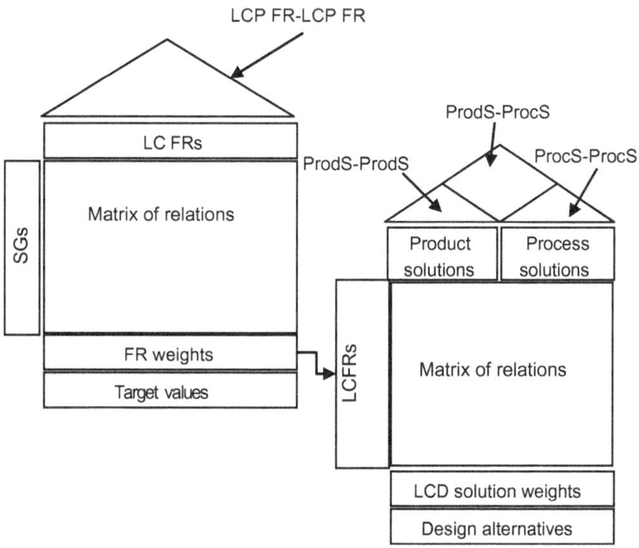

Fig. 6 Modified QFD for integrated design of product and life cycle process

sets of product and process solutions are linked to the LcFRs in the second HoQ. Thus, the correlation matrix in the second HoQ is divided into three matrices: two self-correlation matrices of product (i.e. module, part, component, etc.) and LC process solutions, and one attaching matrix between each product solution with each life cycle process solution. For the relationships between SGs and LcFRs and those between LcFRs and LCD solutions, the relations can be rated on certain types of scale such as (0, 1, 5, 9), where 0 corresponds to no relationship and 9 to strongest relationship. The calculation of the weight of LcFRs, product, and LC process solutions is the same as that of the conventional QFD. By the proposed QFD approach, it can be convenient for the designers to conduct systematic analysis. Designers can quantify the strength of relations between SGs and engineering characteristics of product and its major life cycle processes, as well as those between product and process, i.e. components and its adaptive life cycle options. Therefore, the outputs of the second HoQ are then used to guide the simultaneous generation of product concepts and some feasible life cycle process plans.

2.3 Life Cycle Simulation for Sustainable Value-Driven Product Life Cycle Design Evaluation

Though Life Cycle Assessment (LCA, see Sect. 3.1 for more details) and Life Cycle Costing (LCC) are considered to be powerful tools to assess the environmental impacts and economic viability of a life cycle in a holistic and quantitative manner, they cannot handle the various stochastic factors throughout product life cycle, or execute complicated logics, or evaluate the balance between supply and demand of recycled materials and reusable components in closed-loop-type life cycles. Moreover, LCA and LCC heavily rely on data some of which might be difficult to acquire during the design stage. Instead, Life Cycle Simulation (LCS) as a powerful tool of describing and analysing product life cycles can effectively handle the complexity and innovation in product life cycle design. In particular, LCS makes it possible to estimate the effectiveness of circulation of product, components and materials, as it can simulate the flows of material, energy, information, and cost in product life cycles (Takata and Kimura 2003; Umeda et al. 2006; Matsuyama et al. 2014). By LCS, designers can find out useful information for life cycle design and determine life cycle solutions by constructing models of product and its product life cycle and executing what-if analyses by simulations on the models even with smaller amount of data that might be ambiguous. Though LCS is still under development, it is considered to be a promising supporting tool for life cycle design, especially for the early stage of design.

A product life cycle design evaluation model based on LCS is proposed in this research (see Fig. 7). First, a LCS input model which consists of sub-models of product, life cycle process, and market situations is proposed. It translates life cycle design information and background assumptions into LCS input parameters. The product model contains information of product hierarchical structure, material composition, function deterioration, and designed lifetime. The process model contains information of product life cycle process network which describes the sequential or conditional relations between various unit processes of product life cycle, and the environmental (i.e. materials, energy, emissions) and economic input and output of each unit process. The market model contains information of defined product demands in the market, end consumers' behaviour characteristics (i.e. product use frequency, product load during operation) and preferences (i.e. maintenance preference such as maximum tolerant maintenance cost, minimum tolerant MTBF, etc.).

Second, a general Life Cycle Simulation model of discretely manufactured products is established based on discrete-event system simulation theory (Banks et al. 2005). Elements of product life cycle as a discrete-event system are defined, including "entities" such as product, components, materials, corporates, and end-users, "events" such as "generation of orders on new products", "product arrival at end-users", "product failure during use", "generation of requirements of maintenance", "product retirement", etc., as well as "activities" (or life cycle processes) such as "manufacturing", "use", "maintenance", "collection", "recovery" and "disposal". The event-based simulation strategy is adopted, and the simulation models including an engine model for advancing

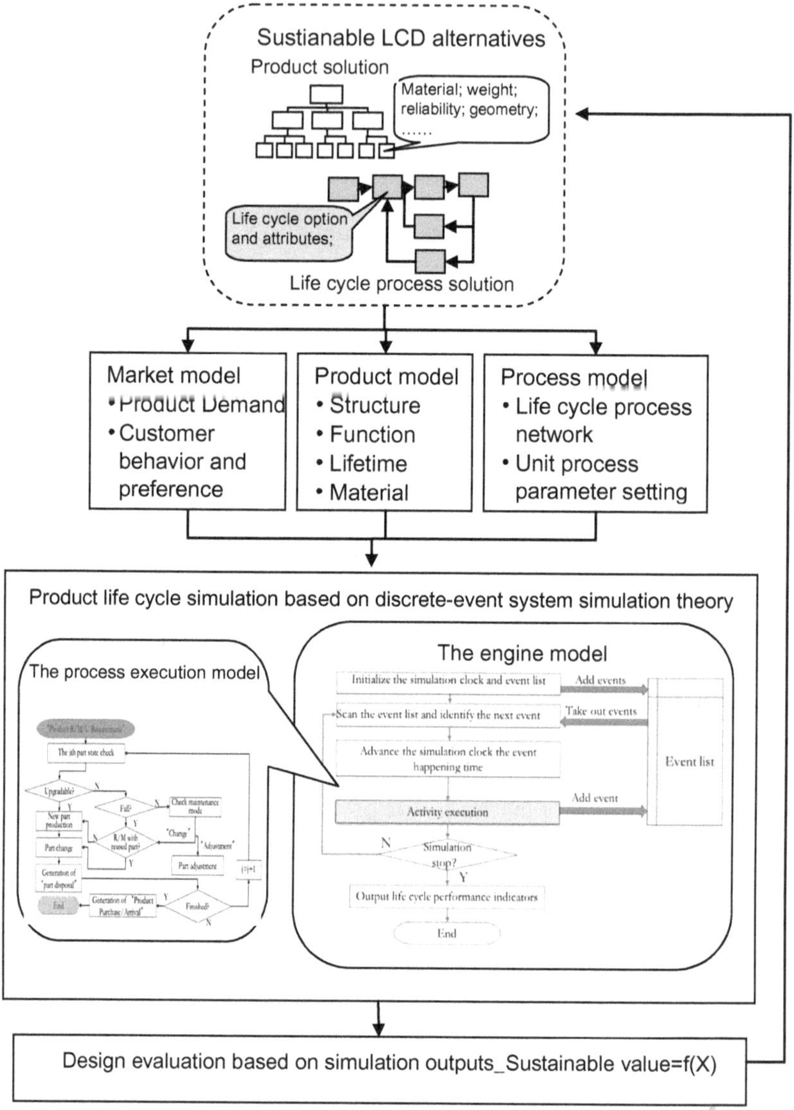

Fig. 7 Product life cycle design evaluation model based on LCS

the simulation, and models for execution of all life cycle "activities" (or processes) are constructed. On start of the simulation, the engine model reads in the event list, finds the event nearest to happen, and then advances the simulation clock to the event happening time and initiates the process which is set to be triggered by the event. The manufacturing process is executed on the order of new products which is either generated according to a given demand pattern or due to retirement of products of end-users. Components can be manufactured using certain ratio of recycled materials provided by the recovery process. Product can be assembled using either newly manufactured components or reusable ones supplied by the recovery process. The ratio of use of the new components to that of the reusable ones is specified according to the life cycle plan. The newly manufactured products are randomly distributed to end-users in need. After the product's arrival at an end-user, the use process is initiated. In the use process, the usage histories of products and components are updated. The usage of products is terminated due to product obsolescence, and interrupted by product failures (which is caused by component failure) or planned maintenance. Occurrence of product obsolescence or failures are determined, respectively, by product value lifetime and component physical lifetime, and the product value lifetime and component physical lifetime are set by random sampling on product purchase and component manufacturing, respectively. The product value lifetime function and component physical lifetime functions are defined in the product model. The products retired either due to obsolescence or being not repairable are then either "collected" or "disposed". The collected used products are then to be recovered. In the recovery process, the products are first disassembled and inspected. The components are either recycled or reconditioned according to the life cycle plan. The maintenance process is executed on requirements of repair or scheduled conditioning. In the maintenance process, the failed components could be replaced by either newly manufactured ones or reusable ones, depending on the life cycle plan. The repaired or conditioned products are then returned to the end-users and continue their usage process. The failed components are either to be recovered or just casually disposed.

During the execution of each process, resource demands, waste amounts, end-user costs, and different corporate costs and revenues are calculated and recorded. The life cycle design alternatives are evaluated based on the simulation outputs which generally include resource and energy demands, waste amount, recovery rate, recycling and reuse rate, end-user costs, corporate costs and profits.

3 CAX–LCA Integration to Support Sustainable Value-Oriented Engineering

With an increasing awareness of environmental crises as well as the growing pressures from the competitors, manufacturers are receiving more concerns with the environmental performance of their products. Life Cycle Assessment is one of the most mature methods to quantify the environmental impacts of product through its life cycle (Baumann and Tillman 2004; ISO 2006). It is of great benefits to perform LCA along with various engineering activities throughout product life cycle, because it is a useful tool predicting the environmental consequences of product and process engineering solutions and therefore can help determine whether if the solution is better for the environment than the currently available ones. However, as comprehensive LCA requires great efforts in data collection, it is very difficult to be conducted during product development, especially in early phases such as conceptual design. Besides, LCA results do not efficiently and explicitly reflect the relations between product and process design factors and product environmental performance, and therefore require professional interpretations to bridge the gap between engineering and LCA domains. For the integration of LCA and product and process engineering, the knowledge needs to be represented in a way that it can efficiently and effectively be shared and used by the engineers from different domains.

Integration of ecological assessment into computer-aided product development environments (i.e. CAX, computer-aided X systems including CAD, CAPP, CAM, PLM, etc.) is proposed so as to potentially ease the difficulties of inclusion of LCA into engineers' daily works (especially the early planning and design phase). Also, feature technology (FT) is considered to be a straightforward approach for realizing data migration between CAX and LCA systems. The feature concept was initially inspired from the desire to support information integration between CAPP (computer-aided process planning) and CAD systems in the manufacturing field. A commonly accepted feature definition is "a generic shape associated with some engineering semantics" (Shan and Mäntylä 1995). To facilitate such integration, the International Organization for Standardization (ISO) delivered the first edition of the standard for "Mechanical product definition for process planning using machining features" as one of the application protocols (APs) of the ISO10303 (industrial automation systems and integration—product data representation and exchange), otherwise known as the standard for exchange of product model data (STEP) (ISO 1994). However, STEP suffers from the rigidity and complexity in implementation and was not intended to share design intents, such as design history and constraints (Xie et al. 2013). Historically, two main research streams are feature recognition and feature-based modelling. Extensive research has been conducted in feature recognition over the past three decades (Vema and Rajotia 2010). The disadvantages of feature recognition algorithms were their complexity and limited types of features that could be recognized (Lam and Wong 2010). On the contrary, another mainstream technological approach is feature-based modelling, which builds models by using feature templates rather than recognizing features from an existing geometrical model. This approach contains rich information associated with design models (Xie et al. 2013). Traditional feature technology in the mechanical design domain has already been well established by many researchers (Babic et al. 2008), and the exploration of this research domain is not the focus of this paper. The concept of feature has been extended and used to bridge mechanical product design and engineering analysis, such as stress analysis with finite element method (FEM) (Lee 2009) as well as manufacturability analysis (Syaimak and Axinte 2011). In the last decade, many researches were proposed to integrate ecological assessment into CAD systems. Otto et al. (2001, 2003) proposed a framework for structured data retrieval in LCA using feature technology and integrating data from a product model and life cycle inventory (LCI) database. Friedrich (1998) tried to integrate LCA to CAD/CAE system and applied it to the product design, which could support the designers to cope with environmental challenges. Nawata and Aoyama (2001) proposed a system especially applicable to machined parts, which automatically generated LCA feedback for the design process. Marosky et al. (2007) presented the structure of an algorithm that allows mutual transfer of data between CAD (SolidEge) and LCA tool (SimaPro). Mathieux et al. (2005) have proposed a tool prototype based on feature technology in extracting CAD/PDM data, from CATIA (CAD) to EIME (LCA). Also, "SolidWorks Sustainability" developed by Dassault Systems allows environmental assessment in real time (once a feature is attributed). However, these works are interested only in the manufacturing phase. In fact, the environmental impacts are generated throughout the life cycle, especially in the use or end-of-life (EOL) phase. However, research suggested that current LCA–CAX solutions are still inaccurate compared to professional LCA tools. This is because: the CAD model represents the final form of design intent, while environmental impacts are estimated from a process perspective; CAD systems do not support LCA data related to processes, machines, purchasing, user and suppliers. Most works of the CAX (i.e. CAD, CAM, CAPP) and LCA system integration are interested only in the manufacturing phase, though its known environmental impacts are generated throughout the whole life cycle. Given the limits of existing studies, the feature-based methodologies to integrate LCA with mechanical product, process engineering are explored in this chapter. The main content of this section include: (1) the feature definition and classification for LCA; (2) the framework of feature-based LCA; (3) the feature-based life cycle modelling approach is developed to address the scheme of representing life cycle processes, life cycle inventory analysis, and enable the exchange of valuable data between the domain of LCA and current computer-aided engineering tools.

3.1 Brief Introduction of LCA

LCA is a tool for quantifying the environmental performance of products taking into account the complete life cycle, starting from the production of raw materials to the final disposal of the products, including material recycling if needed. The leading standards for LCA are ISO 14040: Principles and Framework and ISO 14044: Requirements and Guidelines. ISO 14040 considers the principles and framework for an LCA, while ISO 14044 specifies the requirements and guidelines for carrying out an LCA study. An LCA study consists of four main phases:

Step 1: Defining the goal and scope of the study.
Step 2: Making a model of the product life cycle with all the environmental inputs and outputs. This data collection effort is usually referred to as life cycle inventory (LCI).
Step 3: Understanding the environmental relevance of all the inputs and outputs. This is referred to as life cycle impact assessment (LCIA).
Step 4: The interpretation of the study.

LCA provides the quantitative and scientific basis for all these activities. In many cases, LCA feeds the internal and external discussions and communications. The most important applications for an LCA are:

- Identification of improvement opportunities through identifying environmental hot spots in the life cycle of a product.
- Analysis of the contribution of the life cycle stages to the overall environmental load, usually with the objective of prioritizing improvements on products or processes.
- Comparison between products for internal or external communications, and as a basis for environmental product declarations.
- The basis for standardized metrics and the identification of key performance indicators used in companies for life cycle management and decision support.

3.2 Feature-Based Multi-View Life Cycle Modelling

Feature, in this study, is described as a way to transfer geometric, technological, or functional information of a product entity between various stakeholders throughout the development process and the life cycle. Each feature can be specialized according to its own domain (design, manufacturing, assembly, LCA, etc.). Obviously, every stakeholder focuses on specific information and is only concerned by a set of properties of different entities, which is called the feature view (Bronsvoort and Noort 2004). Therefore, the employment of feature technology for CAX–LCA integration should not only enable the migrants of valuable data between the LCA and current computer-aided engineering tools such as CAD, CAM (computer-aided manufacturing), CAPP and PLM (Product Lifecycle Management) systems, but also to address the stakeholder focus on life cycle processes modelling and life cycle inventory analysis, thus facilitating the development of sustainable products. A feature-based approach is proposed in this chapter to address the problem of life cycle modelling and integrated data management for LCA. However, before going into the details of feature-based life cycle modelling, the definition and classification of features for LCA is presented here first. Generally speaking, the CAX tools can be categorized into types of product-focused such as CAD systems and process-focused such as CAPP and CAM systems. Thus, in order to enable the data exchange between CAX and LCA tools, two types of features which are the product features and operation features are defined in this study, based on engineering domains they are applied to.

- Product features (PFs)

Product feature is defined as an information set for dynamic product representation throughout its life cycle. Product features are then classified into:

- *Form features* form feature refers to a region of a part with some interesting geometric or topological properties. Form features contain both shape information and parametric information of a region of interest. Form features are commonly used as the primary means of creating 3D geometric models in CAD system. Examples of form features include extruded boss, loft, holes, etc.

- *Connectivity features* connectivity feature refers to the relative position and mating relations between parts. Connectivity features can be characterized by the static attributes when the connection has been established or broken, as well as dynamic characteristics about how the connection can be established (during assembly) or can be broken (during disassembly). Thus, by specifying a connectivity feature in a product model, the assembly-specific information known by the feature is also available in the model. In this study, the connectivity feature may contain but not limited to information of: connected product entity; the reference entity for establishing the connectivity; tolerances required to establish the connection; geometric refinements of the connection to ease the assembly operation (e.g. rounds, chamfers, welding grooves).
- *Technical features* technical features are non-geometry/topology-related features, including material features with attributes of material type, material properties, recyclability, status, as well as functionality features with attributes of product entity functionality description and performance parameters (i.e. useful life time, working power, reusability, etc.)
- Operation Features (OFs)

Operation feature is defined as an information set of operations associated with physical and functional changes of product entity, and resulting in environmental impacts. Operation features are characterized by attributes of operation type, operation parameters, and equipment and tooling specifications.

Though product life cycle, in LCA study, is generally modelled as a network of processes such as manufacturing, operation, logistics, use, maintenance, collection, remanufacturing, recycling, and disposal, it can also be defined as a sequence of state changes of different product entities (Riou and Mascle 2009). Thus, a feature-based multi-view life cycle modelling approach is then proposed for integrated life cycle data management. As shown in Fig. 8, each process throughout the life cycle is modelled with views of product state, associated operations and the resulting life cycle inventory. The three life cycle views are integrated based on the feature mapping mechanism between the domains of

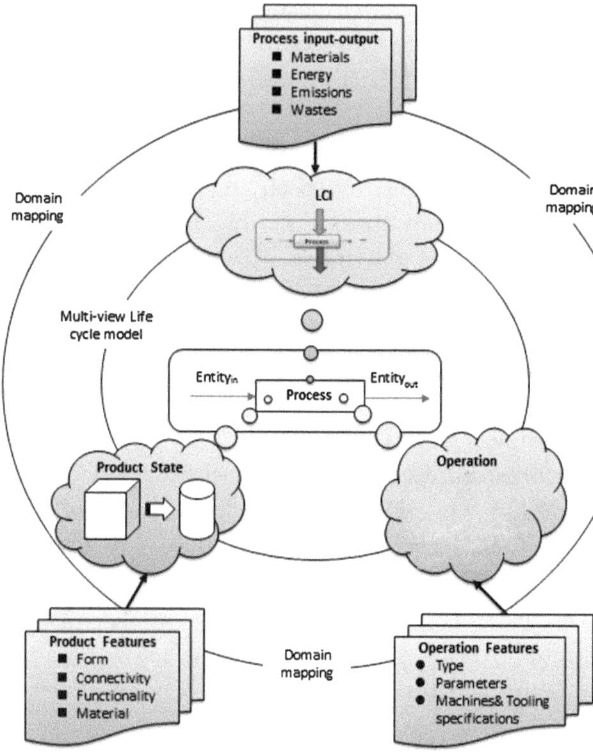

Fig. 8 Feature-based life cycle modelling

product design, process design, and inventory. Thus, the structured relations between various product and process engineering parameters, and indicators of environmental impacts are then developed along with life cycle modelling. The product state view is the dynamic representation of product entities throughout the life cycle. Product state, which may include changes in product shape and form, material properties, functionalities associated with the process is characterized by product feature instances. Operations are defined as the activities causing the product state change and resulting in environmental impacts. The operation view is characterized by operation feature instances. For each process alternative, the operation features can be instantiated by product-to-operation feature mapping algorithms (i.e. STEP—"Standard for the Exchange of Product Model Data"-based process planning based on feature). Based on the product state and operation view modelling, the inventory view is then developed based on the domain mapping form product and operation feature to process inputs and outputs. The process input–output calculation equations should be first established based on process principles and the type of product and operation attributes adapt to the equations are then determined.

3.3 Feature-Based LCA–CAX System Integration to Support Sustainable Value-Oriented Engineering

In this section, the framework for feature-based LCA study is proposed (see Fig. 9). The feature-based life cycle model serves for the data exchange between CAD/CAPP/PLM and LCA systems. Based on the goal and scope definition, the product and process feature attributes required for life cycle modelling and LCI are identified. Necessary data from CAD, CAPP, and PLM systems are collected by feature extraction algorithms. The process inputs and outputs are calculated and then transferred to LCA systems for life cycle inventory analysis and the

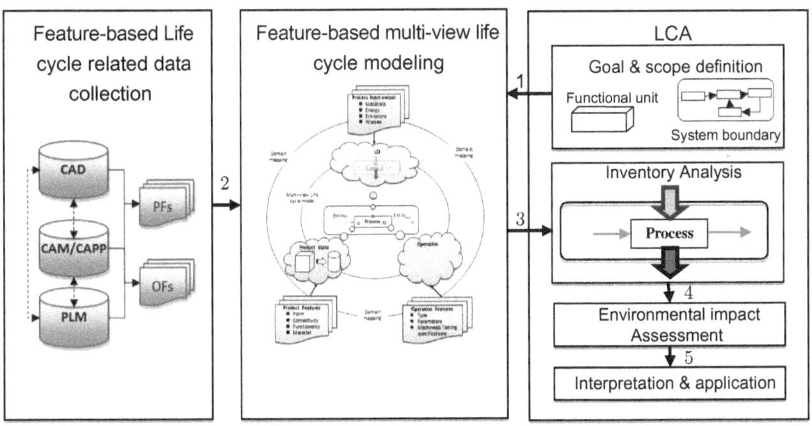

Fig. 9 Framework of feature-based LCA

environmental impact evaluation. Also, by evaluating the sensitivity of impact of product and process feature change on LCA results, the critical engineering factors to environment impacts can be identified.

For the realization of the proposed feature-based LCA approach, the LCA–CAX system architecture is then proposed (Fig. 10). A feature-based LCA tool is proposed for multi-view life cycle modelling and data integration, LCI and impact assessment and improvement analysis. The feature-based data retrievers collect product- and process-related data according to the requirements of life cycle modeller of the LCA tool. The life cycle modeller then generates the multi-view life cycle model and sends the calculated life cycle process input–output data to the LCI and EI calculator. Based on the inventory and impact assessment results from the LCI & EI calculator, the improvement analysis provides modification suggestions based on identification of important feature instances to environmental impacts throughout product life cycle and opportunities for product and process sustainability improvements.

CAD is the most commonly used engineering software for product design. A CAD-LCA software prototype (Tao et al. 2017) to support sustainable value-oriented engineering consists of a plug-in integrator for different CAD systems, a life cycle database and a LCA module for feature-based life cycle modelling and assessment. The plug-in CAD integrator realizes product feature model extraction. The extracted

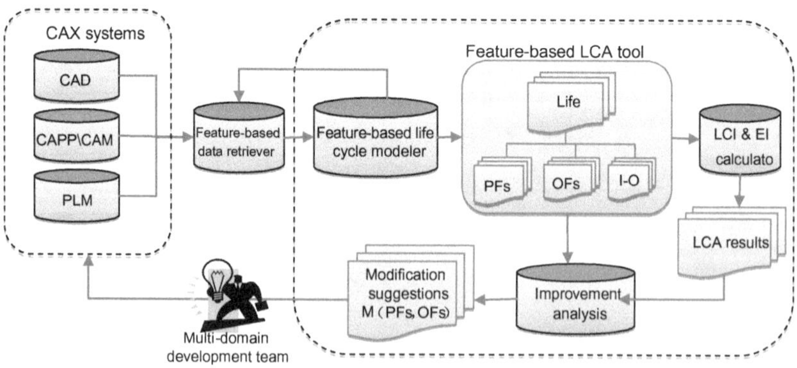

Fig. 10 Feature-based CAX–LCA system architecture

product feature model is sent to the life cycle database with an index tag to distinguish each feature and make a quick search. The LCA system retrieves the product feature model, builds the life cycle process model based on PF-OF mapping, and then calculates the life cycle inventory and environmental impacts based on numerical data from product and operation feature instances.

4 Conclusion

This chapter presents the life cycle design methodologies for sustainable value creation and engineering sustainability in general. The proposed life cycle design framework denotes a systematic and concurrent development of a product and its life cycle process. The key concept of this frame is to organize the integrated product design and process planning as the strategic design mapping process between three domains, including the domain of sustainable goals which are identified sustainable value creation opportunities and potentials in engineering activities throughout the engineering value chain associated with the product life cycle, the domain of product and life cycle process functions, and the domain of life cycle engineering solutions of both product and process. The Life Cycle Simulation is employed for the description of complicated circulation in closed-loop-type life cycle and evaluation of life cycle engineering solutions for sustainable value potentials for different life cycle phases and involved stakeholders. Also, a feature-based CAX–LCA integration approach is proposed for engineering sustainability. It aims at helping engineers to incorporate sustainable value into their daily engineering activities by easier life cycle data retrieval, modelling and environmental assessment. The proposed life cycle design methodologies and tools are expected to help bring experts of product design and life cycle management to share a common vision, that helps avoid conflicts in the traditional design process.

References

Adams, W. M. (2006). *The future of sustainability: Re-thinking environment and development in the twenty-first century.* Report of the IUCN renowned thinkers meeting.

An, Y., Lee, S., & Park, Y. (2008). Development of an integrated product-service roadmap with QFD. *International Journal of Service Industry Management, 19* (5), 621–638.

Babic, B. R., Nesic, N., & Miljkovic, Z. (2008). A review of automated feature recognition with rule-based pattern recognition. *Computers in Industry, 59,* 321–337.

Banks, J., Carson, J., et al. (2005). *Discrete-event system simulation* (4th ed.). New York: Prentice Hall.

Baumann, H., & Tillman, A. M. (2004). *The hitch hiker's guide to LCA.* Stockholm: Studentlitteratur.

Bocken, N. M. P., Short, S. W., Rana, P., et al. (2013). A value mapping tool for sustainable business modelling. *Corporate Governance, 13*(5), 482–497.

Bronsvoort, W. F., & Noort, A. (2004). Multiple-view feature modelling for integral product development. *Computer Aided Design, 36*(10), 929–946.

Cooper, R., & Slagmulder, R. (1997). *Target costing and value engineering.* Oregon: Productivity Press.

Du, Y., Cao, H., Chen, X., et al. (2013). Reuse-oriented redesign method of used products based on axiomatic design theory and QFD. *Journal of Cleaner Production, 39,* 79–86.

Figge, F., & Hahn, T. (2013). Value drivers of corporate eco-efficiency: Management accounting information for the efficient use of environmental resources. *Management Accounting Research, 24*(4), 387–400.

Friedrich, J., & Krasowski, H. (1998). *Ecology-based product data model.* SAE technical paper.

Graede, T., & Allenby, B. (1995). *Industrial ecology.* Englewood Cliffs: Prentice Hall.

Ishii, K. (1995). Life-cycle engineering design. *ASME Journal of Vibration and Acoustics, 117*(B), 42–47.

Ishii, K., & Lee, B. (1996). *Reverse fishbone diagram: A tool in aid of design for product retirement.* Paper presented at the proceedings of ASME design technical conference, irvine, CA,USA.

Ishii, N., Fuchino, T., & Muraki, M. (1997). Life cycle oriented process synthesis at conceptual planning phase. *Computers & Chemical Engineering, 21,* S953–S958.

ISO. (1994). ISO10303-1: Industrial automation systems and integration-product data representation and exchange-part 1: Overview and fundamental principles. Geneva: International Organisation for Standardisation (ISO).

ISO. (2006). *ISO 14040 Environmental management—Life cycle assessment—Principles and framework.* Geneva: International Organization for Standardization.

Kara, S., Mazhar, M., Kaebernick, H., et al. (2005). Determining the reuse potential of components based on life cycle data. *CIRP Annals-Manufacturing Technology, 54*(1), 1–4.

Kato, S., Hata, T., & Kimura, F. (2000). *Decision factors of product life cycle strategies.* Paper presented at the Proceedings of 7th CIRP international seminar on life cycle engineering.

Kato, S., & Kimura, F. (2004). *The product life cycle design method using a strategic analysis.* Paper presented at the proceedings of 11th international CIRP life cycle seminar on product life cycle quality management issues.

Kobayashi, H. (2006). A systematic approach to eco-innovative product design based on life cycle planning. *Advanced Engineering Informatics, 20*(2), 113–125.

Kroll, E., & Hanft, T. A. (1998). Quantitative evaluation of product disassembly for recycling. *Research in Engineering Design-Theory Applications and Concurrent Engineering, 10*(1), 1–14.

Lam, S., & Wong, T. (2010). Recognition of machining features—A hybrid approach. *International Journal of Production Research, 38*(17), 4301–4316.

Lee, S. H. (2009). Feature-based non-manifold modeling system to integrate design and analysis of injection molding products. *Journal of Mechanical Science and Technology, 23,* 1331–1341.

Masui, K., Sakao, T., & Inaba, A. (2001). *Quality function deployment for environment: QFD (1st)—A methodology in early stage of DfE.* Paper presented at the proceedings of eco-design 2001-second international symposium on environmentally conscious design and inverse manufacturing.

Marosky, N., Dose, J., Fleischer, G., et al. (2007). Challenges of data transfer between CAD and LCA software tools. In *Proceedings of 3rd international conference on life cycle management.*

Mathieux., F, Roucoules., L, Lescuyer, L, et al. (2005). Opportunities and challenges for connecting environmental assessment tools and CAD software. In *Proceedings of LCM 2005-innovation by Life cycle management.*

Matsuyama, Y., Matsuno, T., Fukushige, S., et al. (2014). Study of life cycle design focusing on resource balance throughout product life cycles. *Procedia CIRP, 15,* 455–460.

Merante, P., Van Passel, S., & Pacini, C. (2015). Using agro-environmental models to design a sustainable benchmark for the sustainable value method. *Agricultural Systems, 136,* 1–13.

Nawata, S., & Aoyama, T. (2001). Life-cycle design system for machined parts-linkage of LCI data to CAD/CAM data. In *Environmentally conscious design and inverse manufacturing,* 2001. Proceedings *EcoDesign* (pp. 299–302).

Noller, R. M. (1992). Disassembly for disassembly tactioes. *Assembly Automation, 33*(1), 24–26.

Ometto, A, Gueler, A., & Pigosso, D. (2008). *A proposal for a framework for life cycle engineering.* Paper presented at the proceeding of 15th CIPR international conference on life cycle engineering.

Ostad Ahmad Ghorabi, H., & Jerlich, J. (2008). Value based life cycle analysis for occupational. In *Environmental and legal compliance in organizations 1st international conference on industry safety, occupational health and safety in organizations.* Isfahan, Iran.

Otto, H.E., Mueller, K.G., & Kimura, F. (2001). A Framework for structured data retrieval in LCA using feature technology. In *Proceedings of second international symposium on environmentally conscious design and inverse manufacturing* (pp. 250–255).

Otto, H.E., Kimura, F, Mandorli, F., et al. (2003). Integration of CAD models with LCA. In *Proceedings of 3rd international symposium on environmentally conscious design and inverse manufacturing.*

Rana, P, Short, S., & Evans, S. (2013). D2.5—Lessons learned report, documenting the impact from use of the tools and methods and areas for improvement.

Ray, D. L., & Guazzo, L. (1993). *Environmental overkill: Whatever happened to common sense?* New York: Harper Collins.

Riou, A., & Mascle, C. (2009). Assisting designer using feature modeling for lifecycle. *Computer-Aided Design, 41*(12), 1034–1049.

Rose, C. M., Beiter, K. A., & Ishii, K. (1998). *Characterization of product end-of-life strategies to enhance recyclability.* Paper presented at the proceedings of design for manufacturing symposium, ASME, Atlanta, Gergia.

Sakao, T., Masui, K., & Kobayashi, M. (2001). *Quality function deployment for environment: QFDE(2nd report)-verifying the applicability by two case studies.* Paper presented at the proceedings of eco-design 2001—Second international symposium on environmentally conscious Design and inverse manufacturing.

Sakao, T. (2007). A QFD-centered design methodology for environmentally conscious product design. *International Journal of Production Research, 45,* 4143–4162.

SAVE International (Society of American Value Engineers). 2007. *Value Methodology Standard,* US.

Shah, J. J., & Mäntylä, M. (1995). *Parametric and feature-based CAD/CAM.* New York: Wiley.

Suh, N. P. (2001). *Axiomatic design—Advances and applications.* New York: Oxford University Press.

Syaimak, A.S., & Axinte, D. A. (2011). An approach of using primitive feature analysis in manufacturability analysis systems for micro-milling/drilling. *International Journal of Computer Integrated Manufacturing 2011, 22*(8), 727–744.

Takata, S., & Kimura, T. (2003). Life cycle simulation system for life cycle process planning. *CIRP Annals—Manufacturing Technology, 52*(1), 37–40.

Tao, J., Chen, Z., Yu, S., et al. (2017). Integration of life cycle assessment with computer-aided product development by a feature-based approach. *Journal of Cleaner Production, 143,* 1144–1164.

Tchertchian N, Millet D., & El Korchi A (2010). *A method helping to define ecoinnovative systems (product architecture + reverse supply chain structure + use cycles scenario).* Paper presented at the Proceedings of seventeenth CIRP International Conference on Life Cycle Engineering.

Umeda, Y. (2001). Toward a life cycle design guideline for inverse manufacturing. In *Proceedings Second International Symposium on Environmentally Conscious Design and Inverse Manufacturing* (pp. 143–148).

Umeda, Y., Kondoh, S., & Sugino, T. (2006). Analysis of reusability using 'marginal reuse rate'. *CIRP Annals-Manufacturing Technology, 55*(1), 41–44.

Verma, A., & Rajotia, S. (2010). A review of machining feature recognition methodologies. *International Journal of Computer Integrated Manufacturing I, 23*(4), 353–368.

Xie, Y., Liu, J., Liu, H., et al. (2013). Features and interoperability of computer aided engineering systems. *Semantic Modeling and Interoperability in Product and Process Engineering.* London: Springer.

Xing, K., Abhary, K., & Luong, L. (2003). IREDA: An integrated methodology for product recyclability and end-of-life design. *Journal of Sustainable Product Design, 3*(3–4), 149–172.

Yang, M., Vladimirova, D., Rana, P., et al. (2014). Sustainable value analysis tool for value creation. *Asian J. Management Science and Applications, 1*(4), 2014.

Yu, S., Yang, Q., Tao, J., et al. (2011). Product modular design incorporating life cycle issues—Group Genetic Algorithm (GGA) based method. *Journal of Cleaner Production, 19*(9–10), 1016–1032.

Zhang, Y., Wang, H. P., & Zhang, C. (1999). Green QFDII: A life cycle approach for environmentally conscious manufacturing by integrating LCA and LCC into QFD matrices. *International Journal of Production Research, 37,* 1075–1091.

Zhang, L., Zhan, Y., Liu, Z. F., et al. (2011). Development and analysis of design for environment oriented design parameters. *Journal of Cleaner Production, 19*(15), 1723–1733.

Ziout, A., Azab, A., & Atwan, M. (2014). A holistic approach for decision on selection of end-of-life products recovery options. *Journal of Cleaner Production, 65,* 497–516.

Zwolinski, P., Lopez-Ontiveros, M. A., & Brissaud, D. (2006). Integrated design of remanufacturable products based on product profiles. *Journal of Cleaner Production, 14*(15–16), 1333–1345.

Authors' Biography

Jing Tao is an assistant research fellow at School of Mechanical and Power Engineering of Shanghai Jiao Tong University, China. Her research interests include quantified sustainability assessment, and sustainable mechatronics product design and manufacturing.

Suiran Yu is professor of mechanical engineering at Shanghai Jiao Tong University, China. His main research interests include Life Cycle Engineering, and Product Development. He is a senior member of Chinese Mechanical Engineering Society and an overseas member of the Japan Society for Precision Engineering.

13

Engineering and the Skills Crisis in the UK and USA: A Comparative Analysis of Employer-Engaged Education

John R. Bryson, Rachel A. Mulhall, Nichola Lowe and Julianne Stern

1 Introduction

Everyday living in the twenty-first century is underpinned by innovations in the field of *engineering*. Existing products are reengineered and new products created. Engineering underpins innovation in the delivery of services including developments in information and communications technologies (ICT). A successful national economy is one that has a balance between different industrial sectors. This balance includes

J.R. Bryson (✉) · R.A. Mulhall
University of Birmingham, Birmingham, UK
e-mail: J.R.Bryson@bham.ac.uk

R.A. Mulhall
e-mail: rachel.a.mulhall@gmail.com

N. Lowe · J. Stern
University of North Carolina, Chapel Hill, USA
e-mail: nicholajlowe@gmail.com

J. Stern
e-mail: julianne.stern@gmail.com

Y. Zhang and M. Gregory (eds.), *Value Creation through Engineering Excellence*,
https://doi.org/10.1007/978-3-319-56336-7_13

financial services combined with the ability to create high-value-added products and services with a significant engineering element. The Industrial Revolution was founded on engineering innovations, and subsequently periods of rapid economic growth have been closely associated with engineering innovations including steam power, railways, electricity, chemicals, containerisation, logistic management systems, air travels, and computing (Kondratieff 1935). There is thus a significant engineering element that has played a facilitating role in the emergence and ongoing transformation of capitalism. Excellence in engineering is behind innovations in the development and management of infrastructure systems and low- and high-value business-to-consumer and business-to-business products. The term "engineering" has many different meanings. The field of engineering was initially divided into two—civil and military engineers; military engineers specialised in the destruction of works designed by civil engineers. Many different specialist engineering sub-disciplines have emerged including electrical, chemical, and mechanical. There are engineers who create, develop, or invent new products, and then there are skilled engineering-related occupations that are directly involved in the day-to-day fabrication of engineered products.

There are two types of engineered products. First, there are products that are mass produced in factories and, second, special projects that are customised to provide a solution to a particular problem, for example, the construction of a bridge, ship, or building. In 1990, Walter Vincenti, an aeronautical engineer based at Stanford, was asked the question by a colleague in economics "What is it you engineers really do". His answer resulted in a book entitled "*What Engineers Know and How They Know It*" (Vincenti 1990), and part of his answer was "What engineers do, however, depends on what they know". Organising is a key element of any definition of engineering, and Rogers defined engineering as "the practice of organising the design and construction of any artifice which transforms the physical world around us to meet some recognised need" (Rogers 1983: 51). In this context, organise means arranging including the development of new designs whilst "construction" also includes production, fabrication, or manufacture. The Rogers definition is more suited to civil engineering and the development of large projects that require customised construction rather than mass production. Absent from this definition is also anything connected with operational management including maintenance and sales.

The application of engineering practice to manufactured products has increasingly resulted in two outcomes (Bryson and Ronayne 2014; Bryson et al. 2015). First, there is an ongoing escalation in the value created by hi-tech or advanced manufacturing. Low-value manufacturing tasks are either undertaken in lower labour cost locations, or employees involved in the production process have been replaced by machines. High-value-added tasks including research and development have remained in developed market economies (Cohen and Zysman 1987; Bryson and Rusten 2011). Second, there has been an emphasis on the application by engineers of optimisation techniques that have removed raw materials from products and structures. This process has reduced the weight of raw materials embedded in manufactured and constructed products as the design process has focused on efficiency combined with effectiveness (Bryson and Rusten 2011). This shift is part of a transition towards a (Quah 1999; Bryson et al. 2004). The design of structures and products is becoming smarter with a focus on construction and manufacture to finer tolerances to reduce weight and also the cost of raw materials embedded in products or structures (Quah 1999). One consequence of these two outcomes is an increase in the productivity of employees involved in the manufacture or construction of engineered products (Select Committee on Trade and Industry 2002) combined with an escalation in the capabilities or skills required of individual employees (Bryson and Daniels 2008). It is important to distinguish between engineers that design and employees that are involved in constructing and manufacturing engineering products. This chapter focuses on the latter group. A consistent and persistent business and policy concern has revolved around a mantra of skill shortages in manufacturing-related engineering occupations (Leitch Review of Skills 2006; Kumar 2015; UK Commission for Employment and Skills 2016). These skill shortages have the potential to erode the competitiveness of manufacturing in higher labour cost locations. It is noteworthy that such an erosion or decline is unrelated to globalisation and is a purely local affect related to the operation of market imperfections in local labour markets and in educational systems.

Manufacturing-related engineering skill shortages are being experienced in the USA and the UK. In this chapter we compare two models of employer-engagement in the education of 14- to 19-year-olds. The first model, University Technical Colleges (UTC), is supported by the UK government and engages national employers in developing localised solutions to skill demands in science, technology, engineering and mathematics (STEM)-related occupations. The second approach, based in Chicago, USA, is the product of a coalition of manufacturing firms and organisations, the Chicago Manufacturing Renaissance Council (CMRC). The CMRC has supported the development of vocational education in the city via the establishment of the Austin Polytechnic Academy (APA) and a manufacturing-oriented curriculum and training programme within APA called Manufacturing Connect (MC). The APA provides specialist engineering-related education within a "normal" school environment. It pulls in small- and medium-sized enterprises as partner organisations to help inform curriculum design as well as provide work-based learning opportunities for APA students through summer internships, job shadowing, mentoring, and eventual job placement. These two examples offer alternative approaches to developing vocational education for the engineering sector.

2 Skills Gaps, Capabilities, and High Value Engineering

The referendum that was held on Thursday 23 June 2016 to decide whether the UK should leave or remain in the European Union is associated, on the one hand, with considerable political turmoil as 52% of voters stated their desire to leave the EU. On the other hand, the period after the referendum is associated with an attempt to develop a new industrial policy. The driver behind this new industrial policy is closely linked to the outcome of the EU referendum and involves a series of discussions regarding what constitutes an effective industrial policy (Livesey 2012; Clark 2013). Such a policy needs to be long term and must include skills or soft infrastructure, connectivity and related hard infrastructure investments combined with a focus on the wider framework conditions that support economic activity including labour market regulations, the planning system and taxation. A systemic or integrated approach must be developed to create an effective industrial policy.

The history of British industry is one in which manufacturing firms have attempted to overcome some of the constraints imposed upon their activities by their external environment by developing an integrated corporate strategy. For example, from the 1930s, the Smiths Group, formerly Smiths Industries, a British transnational diversified engineering firm with operations in over 50 countries employing around 23,550 staff, developed a skills and housing strategy. In June 1938, Smiths developed an in-house technical school on the understanding that the company's future was "not simply a question of building factories and installing plant—skilled and trained staff were equally vital" (Nye 2014: 97). In the 1950s, the firm found that national government housing programmes were failing to provide sufficient housing to support the company's demand for employees close to their factories. In September 1951, Smiths announced that it was going to develop 150 houses close to one of its factories and another 140 were planned. For Smiths, an increase in their workforce also involved strategies to provide training to enhance employee skills combined with attention to the availability of local housing. It is worth nothing that this firm's strategy was focused on the needs of the business but the policy altered "rather than being content to fund a large portfolio of houses, whether directly or through financing housing associations, 1952 would see a new desire emerge, to sell houses to their occupants wherever possible, to free up cash" (Nye 2014: 143). In this case, the firm was acting as a facilitator in the provision of housing for employees.

There is an ongoing debate in developed market economies regarding manufacturing (Loch, et al. 2007; Bryson et al. 2013; Bryson et al. 2015). On the one hand, manufacturing in relatively high-cost locations is being out-competed by companies producing products in emerging economies with lower factor input costs—land, labour, raw materials.

This has given rise to a well-established debate on deindustrialisation in which employment in manufacturing declines in developed market economies to be replaced by jobs in services (Cairncross 1979; Bryson et al. 2004). The deindustrialisation debate, however, confused a decline in manufacturing employment in response to productivity improvements with an absolute decline in manufacturing (Bazen and Thirlwall, 1991; Bluestone and Harrison 1982; Bryson et al. 2013; Bryson et al. 2015). Employment in manufacturing declined, but at the same time, output increased due to capital investments in machine tools and the application of new technology. The debate on deindustrialisation emphasised the emergence of a new spatial division of manufacturing in which countries with lower labour costs out-competed firms located in high-cost locations (Scott 1986). This is an account based on globalisation and the ongoing development of an international economy. On the other hand, the competitiveness of manufacturing firms located in high-cost locations is constrained by local factors including the availability of appropriately skilled labour and other factor inputs, for example, energy (Mulhall and Bryson 2013, 2014) and land (Kalafsky 2007; Bryson et al. 2013). This is to argue that the competitiveness of manufacturing firms is partly challenged by ongoing developments in the spatial division of labour, but also by local factors or market imperfections that play an important role in undermining the performance, productivity, and growth of manufacturing in developed market economies.

The skill sets required to support manufacturing have altered with the emergence of advanced or hi-tech manufacturing. The UK Leitch report on skills noted that:

> Over the past 20 years or so, the proposition of jobs requiring high skills has increased substantially, as technology and the global economy has changed. Technological change often leads to higher demand for skills and most employers that presented evidence to the Review expressed concern about shortages at intermediate and higher skill levels. Where establishments are undergoing high levels of technological change in their processes, skill needs are reported to have 'gone up a lot' for 42% of jobs, compared to only 25 per cent of jobs in other establishments. These

> changes will continue as the global economy restructures. Only if the UK is a world leader in skills can UK businesses be world leaders in the new global economy (Leitch Review of Skills 2006: 33).

Between 1980 and 2012 per capita world output increased by 1.7% and this corresponds to cumulative growth of just over 60%. This represented a major transformation in lifestyles, consumer behaviours, and labour markets. A relatively modest annual growth rate over a 30-year period involved transformational technological changes including the Internet, mobile computing, smartphones, health care, transport and digital services. According to Piketty:

> These changes have also had a powerful impact on the structure of employment: when output per head increases by 35–50 percent in thirty years that means that a very large fraction – between a quarter and a third – of what is produced today, and therefore between a quarter and a third of occupations and jobs, did not exist thirty years ago (Piketty 2014: 95–96).

The implication of this analysis is that a per capita growth rate of between 1 and 1.8% represents extremely rapid change. For the labour market the rapidity of this change has major consequences for vocational training and for the relationship between skills that exist within a national labour market and forthcoming demand.

In the UK, hard-to-fill vacancies have been a long-term problem. The 2015 employer skill survey of over 91,000 employers found that 6% of all employers had at least one skill-shortage vacancy and that there were "209,500 reported skill-shortage vacancies which was an increase of 43% from the 146,000 reported in 2013" (UK Commission for Employment and Skills 2016: 12). Moreover, since 2013 there had been an increase in skill-shortage vacancies among Machine Operatives (from 25 to 32% of all vacancies) and skilled trades were experiencing the highest density of skill shortages (43%) (UK Commission for Employment and Skills 2016: 13). A review of the state of engineering in the UK in 2015 noted that engineering firms are more likely to experience hard-to-fill vacancies for professionals (31.7%) and skilled trades (24.8%) and consequently "nearly half (48.3%) of engineering enterprises said that hard-to-fill vacancies meant they had delays in developing new products or services whilst 44.8% said they experienced increased operating costs" (Kumar 2015: V). A 2011 joint report by Deloitte Consulting and the Manufacturing Institute echoed similar concerns for the USA, estimating that 2 million highly skilled manufacturing would go unfilled from 2015 to 2025 as a result of sector-wide skill shortages (Deloitte Consulting LLP and The Manufacturing Institute 2011).

There is no question that skill shortages and hard-to-fill vacancies can have a tangible or measureable impact on firm performance including productivity (Christopherson 2012). What is subject to increasing debate, however, is the question of where to place responsibility for persistent skill shortages (Cappelli 2012; Osterman and Weaver 2014; Lowe 2015a). Some labour market analysts have skirted this issue altogether by outright dismissing claims of skill shortages (Sirkin 2011; Salzman 2013). They argue instead that manufacturing wages are set far too low, creating little incentive for skilled, yet underemployed individuals, to change jobs or seek out new employment opportunities that make better use of existing knowledge and expertise. By this same logic, raising wages should help address hard-to-fill vacancies or skill-shortage vacancies by attracting more employees with appropriate skills, qualifications, or experience (Cappelli 2012).

Others have taken a more pragmatic approach, recognising that immediate wage adjustments, in isolation, may not offer a long-term solution, especially as investments in skills must be balanced to reflect both current *and* future needs of local and national labour markets (Glaeser and Saiz 2004; Glaeser et al. 2012). One complicating factor involves an ageing manufacturing workforce, thus putting pressure on companies to recruit younger replacements and by extension, invest more heavily in their skill development. Yet, declining resources within manufacturing firms—itself a reflection of shrinking firm size—means that few but the largest corporations and establishments have the internal capacity to anticipate and prepare for emergent skill shortages (Berger 2013; Osterman and Weaver 2015; Lowe 2015b). External educational institutions are therefore expected to play an active role in addressing this emergent skills challenge. In the right context, they can also be critical to sustained and resilient economic growth. As Piketty argues "the lessons of French and US experience thus points in the same direction. In the long run, the best way to reduce inequalities with respect to labor as well as to increase the average productivity of the labor force and the overall growth of the economy is surely to invest in education" (2014: 307). Still, for educational solutions to be effective, we need to think more critically about what type of educational initiative is needed to support what type of economic actor and activity (Cappelli 2012; Lowe 2015b). Ongoing innovation continues to transform local labour markets implying that forecasting future skill needs is difficult and perhaps impossible for educational institutions to do alone. One option is for employers and educators to forge long-term partnerships to replenish current skill sets whilst also co-investing in future learning opportunities for the next generation of labour market entrants. This replenishment of skills involves employee retraining, but also attracting and training entry-level employees to work in manufacturing.

3 Employer-Engaged Education in England and Chicago

Engineering is suffering from shortages of the "right" entry-level talent. Preparedness for the world of work as well as a practical understanding of engineering is valued by employers and makes school-level graduates more employable. Schools and other providers of secondary education need to be responsive to local employers and engage with them to develop an appropriate curriculum for students. In that capacity, they can do more than simply serve employers the skills they claim they need. They can also mediate exchanges between job seekers and employers in ways that ultimately shapes employer skill demand and who gains access to jobs accordingly (Lowe 2015). This section explores two engineering training skill initiatives both aiming to address skill shortages and career progression for young, entry-level future employees that builds knowledge of the sector and trains students to potentially have future leadership roles in the industry. However, the approach is different in each case.

The national UTC model attempts to build awareness of the sector and a knowledge base needed for high value engineering. In contrast, the Chicago-based model focuses on youth employment within a high-poverty, inner-city neighbourhood with the goal of to delivering workplace ready, accredited skills for immediate entry into the workplace with progression to leadership roles. Both initiatives are responsive to employer needs but at different scales and over different timescales.

3.1 University Technical Colleges, England

The University Technical College (UTC) model was developed by the Baker Dearing Educational Trust (BDET) as a model of vocational education for 14- to 19-year-olds in England. The first UTC was established in 2010, the JCB Academy, with a further 47 opened since and seven further UTCs are due to open over the next 2 years (Baker Dearing Educational Trust, n.d.). A UTC is a specialist school providing a mixed academic and vocational education within the state-funded education system. Each UTC has at least one engineering specialism that is taught alongside academic qualifications within an extended working week and year (up to 40% additional time). Each UTC is supported by employer- and university-partners, a reflection of the desire not to separate vocational and higher education and maintain all progression routes for students. There are two entry points for students: at age 14, which is unusual in the English education system; and at 16, which is a normal transition point after mandatory state education.

The UTC model is a framework for vocational learning. A semi-technical syllabus incorporates vocational learning with traditional academic learning and qualifications. The aim of the schools is to improve the standard of engineering education to meet the needs of high value engineering businesses, as well as provide a progressive career route for young people. This is implemented through project-based learning incorporating traditional academic subject matter within a business and engineering context. The projects are supported in design and delivery by employer-partners. Employer-partners are organisations within the specialist field of the school that are affiliated to the UTC. They provide guidance on teaching content, support delivery of some classes, provide site visits for student groups, and lend their brand to the school to help student recruitment. Employer-partners tend to be large corporate organisations, such as Rolls Royce, JCB, National Grid, and Network Rail, although there are some instances of smaller firms partnering to provide employer-engagement. Each UTC is supported by a network of smaller, local organisations that primarily support student work placements.

The BDET own the model and brand "UTC". This is designed to provide consistency and standards in the vocational learning model across England, with the final aim for the UTC model to become an established route within the English education system. The framework provides an opportunity for nationally significant employers to engage with the initiative nationally (as a member of the BDET national employer panel) or locally with one (or several) UTCs (as a member of the management board or with direct education delivery). The brand provides critical mass providing employers with a unifying framework and parents and prospective students a signal of the quality and durability of educational provision. The UTC model aims to directly address the image problem facing manufacturing in the UK. The perception of a working life in the sector is negative, with careers seen as unstable, limited and dirty compared to more graduate-style service employment. New school buildings, cutting edge technology including the latest machine tools and innovative working practices (9–5 workdays with no homework and open working spaces for students and teachers) help develop a positive image of the future of engineering. This complements more traditional vocational education measures, such as work placements.

Each UTC is designed to reflect the regional industrial structure, through its choice of specialism, as well as to support overall educational provision in the area (supply of good school places). Employer-partners should be based locally and in theory reflects opportunities in the local labour market. However, the primary driver for employers involved in the UTC initiative is to increase awareness and interest in engineering careers rather than in creating a new direct supply of skilled employees to their businesses. The catchment area for the school is sub-regional (which is considerably larger than traditional schools) and is designed to reflect the industrial structure (and consequent skill need) of an area. Despite the growth in number of UTCs over recent years, coverage remains patchy across England. The South West, East Anglia and northern England have relatively few UTCs (two, three, and four, respectively), with the majority located in a central band between London and the North West (Baker Dearing Educational Trust, n.d.).

The UTC model does face some significant challenges. Recruitment is difficult because of the unusual transition point at age 14, logistical challenges of further travel to the schools and the unproven record of UTC performance. This is in part mitigated by the showcase of corporate employers and promise of improved employability of students graduating from UTCs in recruitment drives hosted by employers rather than targeted at local feeder schools. Performance is also mixed. As an indication of performance and standards achieved in the schools, the Office for Standards in Education, Children's Services and Skills (Ofstead) inspection results provide a benchmark for performance and safety across the country in all types of school. Although the initiative is relatively new, 15 of the 48 UTCs currently open have been inspected. The performance of UTCs has been mixed, with only six UTCs receiving a grade of "Good" or above and nine UTCs scoring a grade of requires improvement (7) or inadequate (2) (Ofstead 2016). Three of these UTCs have since closed following poor performance and recruitment issues. There is a tension for these schools between academic and vocational educational performance. Overall, Ofstead inspection reports were positive about the link to employers and equipment for students as it provided a clear progression route, enthused students to self-learn and provided a route to "worthwhile destinations" (Ofstead reports for UTC Plymouth and Liverpool Life Sciences UTC). In comparison, a low expectation of student capability was often cited as a weakness, as well as mixed standards of academic achievement across the schools. Notable was the poor rating of students' literacy skills, even in those schools identified with "Good" performance: literacy, reading, spelling, and grammar are specifically cited as areas of poor performance in 11/15 Ofstead reports.

3.2 Austin Polytechnic Academy, Chicago, USA

Manufacturing Renaissance (MR) was formed in 2005 to address the emerging skill shortage in leadership roles in the manufacturing sector as the ageing workforce began to retire. The MR is a coalition of government, community leaders, small and medium manufacturing companies (SMM), organised labour and education providers. It was considered that there was a need to support the industry by improving entry-level applicants available for employers and promoting leadership skills for young people. The Manufacturing Connect (MC) programme was developed by MR and has been incorporated into the local education system through one school, Austin Polytechnic Academy (APA). APA is funded by the state, but the additional MC activities (equipment and a full-time manufacturing teacher, which equates to around $50,000) are funded by fundraising by MR.

The aim of the MC programme is to address skill shortages in SMMs, but also enhance the education and career opportunities for school-leavers, who live in one of Chicago's most deprived neighbourhoods. MC provides three areas of skill development: technical, work-readiness, and leadership. These skills are developed through work-based learning and contextualised academic learning, as well as specific career-readiness and leadership development courses. In addition, students have the opportunity to acquire accredited metal working skills during the programme.

The MC programme relies on an extensive network of SMMs in the local area to provide students with tours, job shadowing opportunities and postgraduation work placement. Whilst MC has engaged more than 90 SMM in some capacity, a small subset remain active partners and part of a formal advisory group that helps design the programme curriculum and secure equipment donations for training purposes.

Based on the recommendations of partner firms, MC equipment is not leading edge technology. Rather it reflects that currently used by local firms and thus, sets realistic expectations for entering the manufacturing work environment. For the employers, working directly with secondary school students and the education system is a new route to acquiring employees. Traditionally, these firms recruit through word-of-mouth from their workforce. This has provided an informal training mechanism as prospective employees have an awareness of the sector and expectations from existing employees. This recruitment process has become more limited with an ageing workforce, and the MC programme provides an alternative route for recruitment and for students to develop an awareness of employment opportunities in engineering and manufacturing.

The outcomes of the initiatives are mixed. The number of APA students that participate in the MC programme varies between 25 and 75% each year. The factors contributing to low participation are complex, but programme staff acknowledge that APA's location in a high-poverty, high-crime neighbourhood is a contributing factor and where prospective students with strong academic qualifications, including engineering interests, seek out more prestigious educational opportunities outside the neighbourhood. Most MC participants gain some manufacturing work experience during the course of the programme (between 68 and 100%, depending on year between 2011 and 2016). There is greater variance in the number of students that leave with at least one accredited metal working skill (40–100%). The numbers that pursue a career in manufacturing, either through further education or in employment is much smaller, with the average at 20% (varies between 1 and 32%, depending on year). However, leaders of the MC programme do state that this number is higher 1–2 years postgraduation when students return to seeking a job in the sector.

4 Scaling Skills? University Technical Colleges Versus Austin Polytechnic Academy

The skill shortages being experienced by English and American manufacturing firms place limitations or constraints on productivity and growth. The manufacturing workforce of both countries is ageing, and an immediate problem exists with the replacement of retirees from the labour force. This implies that skills initiatives need to respond to both the need to replace retirees combined with demands for higher-skilled employees. The UTC and APA approaches are radically different, but both are relatively insignificant contributions to the skill problem facing high value engineering in the USA and UK. The focus of both initiatives is on hard-to-fill vacancies rather than the skill gap within a firm's incumbent workforce. The difficulty is in shaping the career aspirations of teenagers and in ensuring that there is a relatively direct relationship between educational providers and local employers. Both the UTC and APA go some way to ensuring that there is a direct relationship between some firms and some school students, but the definition of "some" is very restricted. It is also worth considering how a country's skill strategy is shaped by decisions made by relatively uninformed teenagers regarding the focus of their studies rather than shaped or influenced by the needs of a local labour market, by potential employers or government. Thus, millions of incremental decisions made by teenagers and their parents combine together to provide the skill base of a local or national economy and it is this process of incremental decision-making that leads to skill shortages. These decisions are not taken in isolation but are influenced by fashion, peer and parental pressure, individual interest and teachers.

Both employer-engagement initiatives are targeted at educating 14- to 19-year-olds by providing a foundation of contextualised learning that integrates academic with vocational education. Work-readiness and future leadership development are important for employers to meet the changing demands of working in high value engineering industries. However, how the initiatives position manufacturing is very different. An awareness of workplace culture and expectations is a key aspect for both, but the UTC model aims to showcase "corporate" engineering with cutting edge technology to attract new employees into manufacturing. This is based around the provision of best-in-class machine tools that may not yet be industry standard and a focus on understanding the manufacturing sector rather than acquiring a set of directly transferable accredited technical skills.

The APA/MC attempts to illustrate the reality of everyday working practices and experiences for most entry-level manufacturing employees. On the one hand, the APA provides accredited, transferrable skills that are recognised by manufacturing employers. In this case it is possible to argue that APA is more directly engaged with providing potential employees with some manufacturing-related skills compared to the UTCs. The UTCs educate students by providing some understanding of manufacturing, and this reflects a more indirect relationship between the school and potential employees. On the other hand, the APA develops student capability based around using equipment that is used by local manufacturing firms. This equipment includes older equipment that has been donated by manufacturing firms supporting the school. Once again this illustrates the more direct relationship between the APA, students, and the needs of potential local employers.

Both policies operate at different scales, but both attempt to meet the needs of local employers. The UTC model is being applied across England with UTCs being established with the support of large firms that have facilities in local labour markets. The MC programme is a bottom-up intervention that was launched as a single-school initiative, based in high-poverty neighbourhood in a large American city. MC has recently secured a contract with the Chicago public school system to replicate the programme at two other Chicago-based schools. In contrast to the original APA/MC experiment, however, MC staff will work with teachers and students at established high schools, avoiding some of the organisational challenges associated with launching a brand new school. In contrast, the UTCs are a top-down initiative that tries to facilitate or coordinate a local bottom-up solution to the provision of manufacturing-related education across England. This is a novel approach that reflects an interaction between predominantly large firms and an initiative that is designed to raise the profile of manufacturing employment as a potential career option.

There is a tension with the UTC model in that each school has to be evaluated by Ofstead to ensure that they provide an appropriate academic education. The initial assessments have highlighted academic weaknesses in the UTC model primarily because the evaluation emphasises academic rather than vocational criteria. This is not to argue that the UTCs should focus solely on vocational training, but that the Ofstead UTC inspections should acknowledge their distinctive approach as the schools try to balance academic with vocational training.

The difference between the UTC versus APA model is more than scale, but about approach with the APA providing students with recognised metal working qualifications compared to the UTC focus on more standard academic qualifications. There is another difference that is important. Both focus on providing students with an understanding of manufacturing workplace culture to try to ensure that students will consider developing a career in manufacturing. Nevertheless, the UTC focus is on training students in corporate engineering compared to the APA which is more focused on training students for work on the shop floor. For the APA this reflects the needs of local employers, but also the location of the school and its intake in one of Chicago's more disadvantaged communities.

Neither initiative has been around long enough for a rigorous and robust analysis of their local and regional impacts to be undertaken. Education is a long-term process with the evaluation of these initiatives dependent on following complete cohorts through the schools and on into the labour market or further study. It is worth noting that neither the UTCs nor the APA have been extremely successful in feeding students directly into employment in partner manufacturing firms. For many firms their involvement in these local skills-based initiatives appears to be more about corporate social responsibility or corporate philanthropy rather than about training potential employees who can be directly recruited by manufacturing firms. In this case, both the UTCs and APA are intermediate actors in the local labour market as students go on to further study. There is a tension between the ways in which the UTCs and APA position themselves to potential students and their parents as a mechanism for facilitating direct entry to manufacturing employment rather than as an intermediate actor in local labour markets.

5 Conclusions

There is no question that the relationship between available skills in a local labour market and employee need is complex and problematic. Local skill shortages might be overcome by encouraging migration. The Leitch review of skills in the UK notes that "migration generally has a positive effect, helping to mitigate skill shortages and fill jobs that cannot be filled domestically" (2006: 32). Nevertheless, migration as a solution to local or national skill shortages has preoccupied British politicians, Trade Unions, and the media, and this topic is central to the ongoing debate over the EU referendum and the vote to leave the EU. In the UK, across the media and politics, migration has become associated with competition for local jobs and housing and with pressures in local labour markets that are often considered to be problematic. Migration has, however, always played an important role in providing the USA and the UK with access to skilled people.

The skills gap being experienced by US and UK manufacturing firms raises the question of: Whose problem is this and who should be responsible for providing a solution? The solution might be considered to be the responsibility of government, firms, industrial sectors or individuals or some combination. There is a tendency for manufacturing firms to blame government on the assumption that the solution should be developed and paid for from taxation. Alternatively, firms should develop their own solution either working in isolation or in partnership with other firms. There is a blended solution that has been developed in the UK. In the spring of 2017, the way the government finances apprenticeships is altering. On 6 April 2017 an apprenticeship levy will be introduced that requires all employers operating in the UK with a payroll of over £3 million per year to pay 0.5% of their total pay bill as an apprenticeship levy minus an annual levy allowance of £15,000 per year. Firms will then be able to access funding to support the provision of their own apprenticeship schemes. This new system is similar to the UTC approach as it combines a top-down with a bottom-up approach and forces employers to invest in apprenticeship training.

The apprenticeship levy has the scale and the level of inclusion that is absent from the UTC and APA approach. The UTC and APA, however, are playing a very different role in local labour markets—educating students, and their parents, to appreciate or understand high value engineering and the ways in which value is created through engineering excellence. The difficulty is one of scale and scope. The UTC has more scale and the potential for wider rollout across England, but the scope is limited. This limitation is reflected in the emphasis that is placed on understanding manufacturing rather than in providing directly accredited transferable skills. The APA has a very limited scale, but a wider scope as it provides accredited transferable skills. In both cases, the number of students and companies involved is limited and insignificant compared to the scale of the problem. It is possible to argue that the UTC and APA are making a very minor contribution to addressing the skills gap in high value engineering in England and Chicago. Nevertheless, this contribution still matters as it provides an opportunity for transforming the life chances of some individuals and providing some potential employees for high value engineering firms.

Measuring and evaluating the impact of the UTCs and APA requires a longitudinal study based around following cohorts as they enter the schools and then progress on to entering the labour market. Both initiatives require time to become locally embedded. Employer-engagement in the education of young people adds value in the form of exposure to the engineering sector, preparedness for work, providing potential opportunities for employment, and ongoing career progression. This type of value is not identified in traditional measures of academic performance. There is a danger that traditional measures of academic performance undermines or distorts the advantages of including businesses in the classroom. The key questions are perhaps philosophical and revolve around "education for whom and for what?" Is the educational system intended to train or educate citizens for employment or to educate for life? These are not necessarily mutually exclusive propositions. There is reluctance to hardwire the educational system to the needs of local or national labour markets as this may be associated with social engineering to meet employer needs rather than the needs of individual citizens. The answer to these more philosophical questions is one of developing a balance between educating individuals for everyday living and providing labour markets with skilled employees. Part of the answer is based on exploring the tension between academic and vocational training. In this context, the emphasis must be based on developing a balance between different types of skills and qualifications and in ensuring that academic and vocational qualifications are considered to have equal status. An alternative approach is to argue that the changing nature of economic activity means that it is impossible to identify the future skill needs of local labour markets. Cumulative economic growth has transformed labour markets over the last 30 years producing new types of employment. The school system must provide students with a set of adaptable skills and an approach to learning and training that will provide everyone with the capability to respond to radical alterations in the labour market. The only known is perhaps continual radical change in the types of work that will be available over the next 30 years. Individuals, governments, and firms must develop appropriate strategies to respond to these transformations, and this response requires investment in time and training by all involved in local and national labour markets.

References

Baker Dearing Educational Trust (n.d.) *Find a UTC* [online]. Retrieved September 19, 2016, from http://www.utcolleges.org/utcs/.

Bazen, S., & Thirlwall, T. (1991). *Deindustrialisation*. Oxford: Heinemann.

Berger, S. (2013). *Making in America: From innovation to market*. Cambridge: MIT Press.

Bluestone, B., & Harrison, B. (1982). *The deindustrialisation of America*. New York: Basic Books.

Bryson, J. R., Clark, J., & Mulhall, R. (2013). *The competitiveness and evolving geography of British manufacturing: Where is manufacturing tied locally and how might this change?* London: BIS.

Bryson, J. R., & Rusten, G. (2011). *Design economies and the changing world economy: Innovation, production and competitiveness*. London: Routledge.

Bryson, J. R., & Daniels, P. W. (2008). Skills, expertise and innovation in the developing knowledge economy: The case of business and professional services. *International Journal of Services Technology and Management, 9*(3/4), 249–267.

Bryson, J. R., Clark, J., & Vanchan, V. (2015). *Handbook of manufacturing in the world economy*. Cheltenham: Edward Elgar.

Bryson, J. R., Daniels, P. W., & Warf, B. (2004). *Service worlds: People, Organisations, Technologies*. London: Routledge.

Bryson, J. R., & Ronayne, M. (2014). Manufacturing carpets and technical textiles: Routines, resources, capabilities, adaptation, innovation and the evolution of the British textile industry. *Cambridge Journal of Regions, Society and Economy, 7,* 471–488.

Cairncross, A. (1979). What is de-industrialisation? In F. Blackaby (Ed.), *De-industrialisation* (pp. 5–17). London: Heinemann Educational Books.

Cappelli, P. (2012). *Why good people can't get jobs.* Philadelphia: Wharton Digital Press.

Christopherson, S. (2012). Job creation strategies to accelerate the return of US manufacturing. *Progressive Planning, 190,* 13–16.

Clark, J. (2013). *Working regions: Reconnection innovation and production in the knowledge economy.* New York: Routledge.

Cohen, S., & Zysman, J. (1987). *Manufacturing matters: The myth of the post-industrial economy.* London: Basic Books.

Deloitte Consulting LLP and The Manufacturing Institute (2011). *Boiling point? The skills gap in U.S. manufacturing.* Retrieved February 13, 2017, from https://www2.deloitte.com/us/en/pages/manufacturing/articles/boiling-point-the-skills-gap-in-us-manufacturing.html.

Glaeser, E. L., & Saiz, A. (2004). The rise of the skilled city. *Brookings-Wharton Papers on Urban Affairs, 5,* 47–94.

Glaeser, E. L., Ponzetto, G. A. M., & Tobio, K. (2012). Cities skills and regional change. *Regional Studies, 48*(1), 7–43.

Kalafsky, R. V. (2007). An examination of the challenges of charlotte manufacturers. *The Professional Geographer, 59*(3), 334–343.

Kondratieff, N. (1935). The long waves in economic life. *Review of Economic Statistics, 17*(6), 101–115.

Kumar, A., Randerson, N., & Johnson, E. (2015). *Engineering UK 2015: The state of engineering.* London: Engineering UK.

Leitch Review of Skills. (2006). *Prosperity for all in the global economy—World class skills.* Norwich: HMSO.

Livesey, F. (2012). Rationales for industrial policy based on industrial maturity. *Journal of Industry, Competition and Trade, 12*(3), 349–363.

Loch, C. H., Chick, S., & Huchzermeier, A. (2007). Can European manufacturing companies compete? Industrial competitiveness, employment and growth in Europe. *European Management Journal, 25*(4), 251–265.

Lowe, N. (2015a). From skills mismatch to reinterpretation: Challenges and solutions for manufacturing worker retention and recruitment. In J. Bryson,

J. Clark, & V. Vanchan (Eds.), *Handbook of manufacturing industries in the world economy*. Cheltenham: Edward Elgar.

Lowe, N. (2015b). *Reskilling urban manufacturing: Workforce solutions from the inside out. Metropolitics*. Retrieved December 15, 2015 from http://www.metropolitiques.eu/Reskilling-Urban-Manufacturing.html.

Mulhall, R., & Bryson, J. R. (2014). Energy price risk and the sustainability of demand side supply chains. *Applied Energy, 123,* 327–334.

Mulhall, R. A., & Bryson, J. R. (2013). The energy hot potato and governance of value chains: Power, risk, and organizational adjustment in intermediate manufacturing firms. *Economic Geography, 89*(4), 395–419.

Nye, J. (2014). *A long time in making: The history of smiths*. Oxford: Oxford University Press.

Ofstead. (2016). *Inspection reports* [online]. Retrieved September 19, 2016, from http://www.ofsted.gov.uk/inspection-reports/find-inspection-report.

Osterman, P., & Weaver, A. (2014). Skills and skill gaps in manufacturing. In R. Locke (Ed.), *Production in the Innovation Economy*. Cambridge: Massachusetts Institute of Technology Press.

Piketty, T. (2014). *Capital in the twenty-first century*. Cambridge MA: Harvard University Press.

Quah, D. (1999) *The weightless economy in economic development* (Centre for economic performance, Discussion Paper No 417). London School of Economics.

Rogers, G. F. C. (1983). *The nature of engineering: A philosophy of technology*. London: Macmillan.

Salzman, H. (2013). What shortages? The real evidence about the STEM workforce. *Issues in Science and Technology, 29*(4), 58–67.

Scott, A. (1986). Industrial organization and location: Division of labor, the firm and spatial process. *Economic Geography, 62,* 215–231.

Select Committee on Trade and Industry. (2002). *The competitiveness and productivity of UK manufacturing industry*. London: The Stationery Office.

Sirkin, H. L., Zinser, M., & Hohner, D. (2011). *Made in America again: Why manufacturing will return to the U.S.* Boston: The Boston Consulting Group.

UK Commission for Employment and Skills. (2016). *Employer skills survey 2015: UK results*. Rotherham: UKCES.

Vincenti, W. G. (1990). *What engineers know and how they know it: Analytical studies from aeronautical history*. Baltimore: The Johns Hopkins University Press.

Authors' Biography

John Bryson is Professor of Enterprise and Competitiveness, City-Region Economic Development Institute, University of Birmingham, UK. His research focusses on exploring the competitiveness of firms and city- regions with a focus on knowledge, learning, services and skills.

Rachel Mulhall is an Honorary Research Fellow at the Birmingham Business School, University of Birmingham. Her research focuses on regional economic development, local forms of competitive advantage and the ongoing development of the manufacturing sector in the UK.

Nichola Lowe is associate professor in City and Regional Planning at the University of North Carolina, Chapel Hill. Her work focuses on the institutional arrangements that lead to more inclusive forms of economic development and especially efforts to align growth and equity goals through strategies of business engagement.

Julie Stern is a Housing Development Officer with the City and County of Denver, where she assists affordable housing developers in securing public investment. As a Masters of City and Regional Planning student at the University of North Carolina at Chapel Hill, she collaborated with Dr. Nichola Loweto research how sector-based workforce training initiatives relate to employer partners.

14

An Industrial Policy Framework for High Value Engineering

Chris Collinge

1 The Context

In this chapter I argue that industrial policy obeys a logic, and I suggest that this logic—as it bears upon different sectors including high value engineering—can be analysed in terms of the relationship between business and its environment, including particularly the governmental environment. The relationship between businesses and the state is a reciprocal one, with businesses impacting upon the state in various ways, and the state in turn shaping the structure of the economy and the environment for business development. However, when we focus upon the impact of the state upon the economy, and upon particular sectors within the economy, then it emerges that this impact can take only a limited variety of forms, at least within liberal market or coordinated market economies (Hall and Soskice 2001). By sketching a general

C. Collinge (✉)
Birmingham Business School, Birmingham University, Birmingham, UK
e-mail: C.J.Collinge@bham.ac.uk

Y. Zhang and M. Gregory (eds.), *Value Creation through Engineering Excellence*,
https://doi.org/10.1007/978-3-319-56336-7_14

framework for analysing industrial policy options, we can identify the policy options available to government in the promotion of particular sectors such as high value engineering, and we can also identify the choices that have actually been made regarding different sectors in the past and the reasons for those choices. To begin with, the framework that is proposed here is necessarily static, but subsequently, it is possible to place this within a dynamic context by accommodating certain variations in the performance of economies over time and space. In the second part of the chapter I therefore consider how the industrial policies pursued by government have varied, in response to unstable and uneven development, and how this has been affected by changes to governance patterns.

2 The Logic of Economic Policy

Historically, there has been a close bond of mutual influence between the state and the economy. The economy in a particular place, in its structure and its behaviour, has had a major impact upon the organisation, scope, and priorities of the governing institutions and authorities in that place. Authorities in places with strong manufacturing industries, for example, have had to reflect the interests of these industries in their services and have derived both their incomes and their personnel (politicians and officials) from the communities that depend upon these sectors. As the importance of high value engineering has grown around the world, so the impact of this sector upon the composition and behaviour of state agencies in particular places has grown correspondingly. At the same time, however, governing authorities have also shaped the economy in its development at national and sub-national levels, through their general activities as well as through those policies that are specifically designed to be 'economic'. Governing authorities are amongst the largest employers in most areas, for example, and they have contributed significantly over the years to the physical and social infrastructures (utilities, housing, education, social services, leisure, cleansing, and environment) that make private enterprise viable and successful. Alongside these general activities, however, they have also developed specific industrial policy instruments to influence economic outcomes.

3 The Statics of Industrial Policy

Economic policies can take a wide variety of forms according to circumstances and the interests of those involved in their formulation and implementation. To clarify the policy choices that have been made by national and sub-national agencies, and the ways in which these choices have been constrained by intra- and extra-governmental pressures, it is necessary to begin by identifying the different types of economic initiative that are in principle available for implementation. This framework applies primarily to those societies where a division has been instituted between the state and the economy, between the 'public' and 'private' sectors, whether they are liberal market or coordinated market economies. It applies less clearly if at all to those remaining societies in which no such division between the state and the economy has been instituted, or in which this division is emerging slowly. There are many different ways of analysing and classifying economic policies and activities. For our purposes here it is important to suggest a way into this type of analysis.

3.1 Basic Approaches to Economic Policy

The economic system is made up of processes (transactions such as managing, cooperating, exchanging, and competing) which take place between participants (employers, employees, firms, governments, and other agencies) who have different amounts of power. However, before analysing these instruments in detail it is possible to distinguish three broad orientations towards economic policy on the part of government, orientations that frame their policy choices:

- *the active approach*, whereby government (including sub-national government) uses positively the powers and instruments at its disposal (such as the ability to set interest rates or tax rates, and to build infrastructure) to influence an economic process (such as the attraction or retention of innovative businesses) as a way of achieving a particular end result (such as economic growth, productivity growth, job creation, an improved trade balance). This approach is based upon

the belief that the state must act upon the processes concerned because otherwise the desired result will not be achieved, or achieved as well.
- *the counter-active approach*, whereby government reduces its use of the powers and instruments at its disposal, and acts assertively to limit the powers at the disposal of other bodies (such as at other level authorities, private cartels, or trade unions) to influence economic processes (such as FDI, company recruitment practices), and thereby releases market forces from prior intervention as a way of achieving a particular end result. This approach is based upon the belief that government and other powerful agents must act less upon the processes concerned—for example, they should not attempt to 'pick winners'—and that by doing so they leave decisions to other agents that are better placed to decide.
- *inactive approach*, whereby government does not use the powers and instruments at its disposal to intervene in an economic process, but neither does it act to limit the power at the disposal of other bodies (such as, say, monopolistic private utilities) that already influence the processes concerned. This approach is based either upon an ignorance of particular economic issues, or upon political paralysis or grid-lock, or upon a belief that the current mixture of market and institutional forces is about right already and producing the desired result. Politics is, after all, the art of the possible.

Each of these approaches has been operative over the years across capitalist economies, and each has been adopted on occasions by national and sub-national governments. The first two may be regarded as all-embracing philosophies, perhaps labelled 'interventionist' and 'disinterventionist', which determine most details of policy and define opposite ends of a political spectrum from left to right. Coordinated market economies may lean more towards the interventionist end, and liberal market economies towards the disinterventionist. But in practice it is likely that all three approaches will be combined, with active policies towards some parts of the economy (e.g. the politically sensitive agricultural or financial sectors) being pursued alongside counter-active policies towards other parts (e.g. less favoured manufacturing industries), and inactive policies elsewhere. This is very likely in federal systems such as the USA, where power is divided and dispersed. Furthermore, it is likely at least in European countries that counter-active policies (such as efforts to reduce 'red tape' and 'bureaucracy') can only be taken so far, and that beyond this point there will be social, political, and economic pressures to retain existing arrangements (such as the legal and administrative framework protecting health and safety at work, or the development of an economic or social crisis on the back of excessive de-regulation, as in the case of financial liberalisation). The debate between political parties in practice concerns where the balance between these three dispositions is to be struck in regard to each part of the economy.

3.2 The Range of Policy Instruments

To explore further the options open to industrial policymakers, however, it is necessary to examine the powers and instruments which states and governments have at their disposal, powers which they may choose to use actively, or to renounce using for counter-active reasons. The instruments of economic policy are those factors that are under the control of the policymaker and that can be altered in order to achieve a desired objective. The objective might be, for instance, to attract or to expand high value engineering businesses. The instrument in this case could be assistance with inward investment or capital formation, with R&D, training, and apprenticeships.

There are two broad types of contribution which the different tiers of government in market economies have made over the years to economic development. First of all, they have taken responsibility for providing a social framework (e.g. social order, cultural orientation, legal and property rights, housing and health care) within which private enterprises can flourish. Secondly, they have also taken responsibility on many occasions for producing the economic conditions that permit and encourage enterprise (such as an institutional framework of trade agreements and regulatory equivalence, macroeconomic fiscal and monetary policies, microeconomic policies for particular sectors, appropriate infrastructure, an educated and skilled workforce, access to information and ideas, social capital in the form of networks and clusters, and sometimes risk capital). It is the second of these two contributions that is normally regarded as 'economic policy', and this has been performed in a variety of different ways, sometimes by supplementing or overriding the market, sometimes by removing impediments to the operation of market forces.

Much economic policymaking (although by no means all) is designed to influence the performance and behaviour of employers—usually private but also public and voluntary sector organisations. The existence of businesses and other employing organisations is based upon certain preconditions that must be established and maintained. These preconditions include an ordered social framework (involving, for instance, contract law, dispute settlement, or adequate public and environmental health). They also include the ability to draw productive resources into the enterprise, and to deliver finished goods and services back out into the economy. It is possible to regard each employing organisation as the centre of a web of connections comprising links on the input side of production, links on the output side, and links to the wider social framework, which are listed in the Table 1.

It is through these connections that government acquires the levers or instruments of economic policy. Clearly, these instruments cannot be operated in isolation from one another but are interdependent (e.g. providing industrial accommodation creates a demand for building

Table 1 The instruments of economic development

Social framework	Defence, property rights, law making and enforcement, cultural dispositions (e.g. entrepreneurialism, trust), housing, education, social security, family stability`
Physical infrastructure	Utilities (water, power, communications), commercial land and premises (e.g. science parks), highways, communications including transport networks and fibre optic networks
Factors of production	Plant, equipment, finance (debt, equity, grants, subsidies), technological knowledge, business information and advice, skills—in business, research and design, making, selling,
Factors of consumption	To boost demand in the economy, by producers and consumers, by the public or private sectors
Social capital	Vertical and horizontal collaborative relationships (e.g. 'triple-helix' brokerage) building and sustaining high-trust environments

services, and may draw firms into the area that will create a market for other businesses). In practice, different elements have been grouped together—whether for political, technical, or administrative reasons—forming policy patterns or modalities such as 'public enterprise' or 'industrial development'. The purpose here is to show how the different policy patterns that are acknowledged by practitioners are composed by drawing upon the range of instruments that are in principle available. Each category will be examined in more detail.

3.3 Social Framework

The disruptive economic effects of social disorder are especially apparent at certain times in history (civil war, organised crime, family breakdown, and anti-social behaviour). Governing authorities have contributed over the years to the development of the economy by establishing a degree of social order based upon policing, social security systems (which serve in part to pre-empt the disorder that would result from widespread destitution), medical services and environmental health provision (given the disrupting effects of mental illness, infectious diseases and epidemics, and uncollected waste). The legislative responsibilities of central government have established property rights, civil and criminal laws (commercial law, for instance, defines corporate status and limited liability), regulations that set minimum standards of business behaviour (e.g. regarding employment practices), criminal law and the prevention of corruption. More recently, authorities have contributed to an ordered community by providing housing and education services, and by supporting family stability.

3.4 Physical Infrastructure

A major contribution of authorities to the economy over the years has concerned physical infrastructure such as the provision of utilities, land and buildings. However, the nature of this contribution has necessarily taken different forms at different times. During the early phase of industrialisation in Europe, local and regional authorities undertake the provision of utilities—water, gas and electricity, transport—through a pattern of trading activity. During the Edwardian period in the UK local councils became indirectly involved in the provision of land by using the regulative powers available for town planning. In the 1920s and 1930s, and once again in the 1960s, British local authorities began to construct industrial sites and premises, and to promote these to attract mobile industries by way of industrial development committees and boards. They also developed tourist attractions and promoted these. In some cases local authorities have provided financial assistance—grants or loans or subsidies such as rent or rate relief attached to the offer of sites and buildings. In the 1980s and subsequently science parks and incubator units have continued to be built, as well as *grands projects* like Trans-European Networks and the Channel Tunnel.

3.5 Factors of Production

During the 1970s there was growing awareness on the part of government in Europe and the USA of the non-physical requirements of business, and in the 1980s some governing authorities began to provide specialist advisory services regarding particular issues (such as marketing) or the needs of particular sectors (such as motor vehicles or ethnic minority businesses). Here we see a decisive move towards the pattern of economic policy known as economic development. In some cases regional authorities developed light industrial units adjacent to the science faculty of the university, creating a 'science park' that was intended to enable technology to be transferred from research into business.

It has been argued repeatedly that enterprising companies are being starved of finance. In particular, it is argued that the clusters of business activities which thrive in certain areas depend upon a second tier of venture capitalists and consultants who facilitate the flow of resources (Kenney and Von Burg 1999). Without necessarily using their own funds, governing authorities can and have addressed this problem by bringing venture capitalists into their areas and arranging surgeries for these with entrepreneurs. Many authorities have also supported training schemes as part of an economic development programme, thereby helping to enhance the employability of unemployed people.

3.6 Factors of Consumption

National and regional agencies have sometimes used their purchasing power to supplement demand in the economy, diverting their own spending where possible towards local firms. This is very apparent in the use of defence contracts, for example, and high value engineering will benefit from high speed rail, nuclear power, renewable energy, and other large public projects. In the 1980s many local authorities sought to persuade companies to contribute to employment by making contracts for the purchase of goods or services conditional upon them doing so. For a century or more government at all levels have also attempted to bring demand into the area by attracting tourists, and this activity has taken off in recent years with municipal investment in major schemes that will appeal to visitors. At the same time authorities have recognised the importance of international tourism and now organise overseas promotions and business conferences. Inward investment can be attracted by creating the right macroeconomic and trading conditions and will be discouraged where these conditions are unhelpful.

3.7 Social Capital

The willingness of businesses within an area to link up with one another in supply/value chain relationships, partnerships, or collaborations of various sorts depends upon a flow of information against a background of networking experience and more particularly of mutual trust. It is generally recognised that trust represents a much more efficient way of handling transaction costs than contracts involving lawyers (Fukuyama 1995). Authorities may contribute to this by bringing potential collaborators together from (say) universities, enterprises, and venture capital or other sponsoring organisations in a manner that relates to what has been described as the triple helix (Etzkowitz and Leydesdorff 2000).

3.8 The Targeting of Policies

It is likely that the instruments of policy which have been identified above will be targeted, that they will be applied selectively within the economy to particular types of employing organisation. There are as many different ways of selecting targets as there are ways of classifying employers, but in practice authorities have used criteria such as:

- where employers are located (e.g. in the inner-city),
- company size or age (e.g. small firms, new firms, large overseas firms),
- sector (e.g. high value engineering; business services; motor vehicles; electronics) or cluster (e.g. life sciences or biotech),
- company form (e.g. private, mutual society, public limited, cooperative, community enterprise),
- race or gender of proprietors (e.g. ethnic minority business).

Business advice, for instance, has been tailored to the needs of cooperatives or of ethnic minority businesses, and whilst some governments have constructed whole strategies around assistance to the private sector others have focused upon the public sector or cooperative and community enterprise development.

4 The Dynamics of Industrial Policy

The next step in the development of this framework is to move from presenting a static homogeneous picture of industrial policy towards one that allows for dynamic change and heterogeneity. Two particular issues with which governments at every level are repeatedly preoccupied are the instability of economic progress over time from one year to the next and the unevenness of economic progress over space, from one place to another. Each of these issues has posed difficult problems for government to resolve.

4.1 Unstable and Uneven Development

Most economies experience fluctuations in activity over time, and over a period of perhaps 8 or 10 years it is generally possible to observe a cycle of business expansion and contraction. Typically, the business cycle involves the growth of activity to a point where further expansion is constrained by structural barriers of one sort or another (such as the quantity of available capital or labour) at which point inflation or balance of payments difficulties cause activity to slow down or to contract. Less typically, but more importantly since the 1980s, instability has also been associated with speculative investment bubbles fuelled by over-exuberance producing a series of Minsky Moments in which asset prices collapse and indebtedness rises to unsustainable levels. The regular business cycle is therefore now combined with a more erratic cycle of speculative bubbles and credit crunches, affecting different sectors. In 2007, for example, we saw the culmination of such a speculative housing bubble facilitated by low interest rates and the securitisation of mortgages in the USA, followed by a classic Minsky Moment in which the international banking system froze and nearly collapsed, and sovereign debt rose to new highs. This required massive government intervention including taking-over banks and insuring private savings.

In Britain, Europe, and the USA there has also been a tendency over the years for economies to develop unevenly from one place to another, with some nations, regions, and districts growing ahead of the whole, whilst others lag behind to greater or lesser degrees. This unevenness reflects differences in the conditions which are encountered by businesses in different places. Locations vary, for instance, in the access they offer to raw materials, skills, sources of finance, and markets for finished goods and services. There is often unevenness at the sub-national level between faster- and slower-growing regions, and the composition of the economy can be an important factor here. Jane Jacobs has, for example, stressed the virtues of each region or city containing a related variety of sectors—not too homogeneous, not too heterogeneous—as a basis for sustained growth (Jacobs 1972; Cooke and Leydesdorff 2006). There is also unevenness within regions between inner-city, suburban, and rural areas. Inner-cities and urban areas even within prosperous regions may have depressed conditions compared to surrounding small towns, suburbs, and rural areas. It has been argued that this reflects the constraints upon living and working in these areas, the differential availability of entrepreneurial skills, or of suitable sites for expansion or modernisation.

The pattern of unevenness across a nation evolves over time as the constraints upon business activity change, and as companies have encountered new patterns of demand in product markets. Improvements in electricity supply, transportation, computers, and telecommunications, for instance, have given firms more choice about location, allowing them to move nearer to consumers, or closer to administrative and governmental centres, or into areas where labour is cheaper and the environment more congenial. These changes have also given larger multi-plant firms the flexibility to establish a spatial division of labour, to move their activities from one area to another as conditions fluctuate, and to respond to recession by withdrawing investment altogether from those places where production is more costly or less efficient. These processes have important consequences for the well-being of communities, allowing the physical environment and standard of living to fall below the national average in depressed areas, causing problems of overheating and congestion in areas that are growing too quickly (Harvey 2007). It also has implications for the well-being of businesses, allowing some to benefit from external economies, whilst others fail to do so, or suffer adverse external economies. There is also a connection between instability and unevenness. When the economy is in a downturn phase it is often (but not always) the areas which are already depressed that experience the greatest contraction in activity, and those that are affluent which experience the least.

4.2 Centralisation, Fragmentation, and Decentralisation

Since the early 1930s there has been a continuous centralisation of control over sub-national authorities in the UK, over their structures, resources, operations, and functions, and this process has modified the relationships between authorities and their economies by redirecting these through the apparatus of national or supra-national government. One of the purposes—and benefits—of this centralisation has been to compensate sub-national authorities for economic unevenness, dampening the fiscal impact of economic change by drawing upon resources that have been pooled at the centre. From the point of view of national or federal government, another benefit has been the ability to target redevelopment resources where they are needed most, or will do the most good, and at the same time to draw sub-national expenditures into the framework of national fiscal and monetary policies.

One of the drawbacks of centralisation, however, is the danger of diminishing the responsiveness of government to the needs and interests of businesses and communities. It is to reduce this danger that the influence of national or supra-national government has over the years been channelled through centrally appointed regional offices, or regional development agencies (Collinge and Srbljanin 2002). Indeed, reviews of sub-national economic development and regeneration by national government have pointed towards a greater role for sub-national agencies in the development and delivery of economic policies. So although it is now mediated via a multi-scalar hierarchy, the interdependence of governments and economies continues to the present day, with sub-national agencies dependent upon their economies, and businesses requiring government assistance from one level or another. Indeed, by increasing the role of the market in government functions, and by strengthening the role of partnership in the provision of services, central government has in many ways reinforced the integration and interdependence of governance and the economy.

4.3 European Integration and Globalisation

These processes of centralisation must themselves be placed in a wider geopolitical context by observing that they do not end at the national borders. A major factor has been European integration which continues around the euro despite recent growth in protectionist and nationalistic tendencies. A similar increase in the exposure of places—economies and authorities—to external influences has been occurring under the process of globalisation. The UK government has been a particular champion of globalisation since the 1980s, as can be seen in the restructuring of the City of London that occurred under the heading of the Big Bang in 1986. During the Big Bang a restructuring of the London stock and bond markets was combined with the introduction of a new light-touch approach to regulation, and an openness to foreign banks and to the concentration of finance capital internationally. At the same time many building societies were demutualised and began to adopt a more aggressively speculative approach to borrowing and lending. All of this itself took place against a background of increased international capital flows and the removal of national restrictions upon these flows that was overseen by the World Trade Organisation (Epstein 2005). The effect was to strengthen the competitive position of the City of London during the 1990s, turning it into the world's most important banking centre, and paving the way for a long period of great prosperity for city workers, for parts of Greater London, and for the wider region of the south east. At the same time, through its focus upon investment in international equity and currency markets, and more recently upon the housing market and upon mortgage-backed securities, this process did little to support domestic activity in midland and northern regions, and may even have harmed manufacturing through its impact upon the sterling exchange rate.

4.4 Growth Management and Growth Promotion

As regards overtly economic policies, for over a century governing authorities in developed economies have been devising ways both to manage economic growth and to smooth the process of economic transformation or decline. During periods of economic growth, authorities must manage this growth and the demands that it imposes upon overheating markets—labour and capital, land, infrastructure and property markets—through the various levers at their disposal. Such *growth management* strategies will (depending upon circumstances) use levers that include the expansion of education and housing provision, the expansion of development control, and the proactive investment in office and factory accommodation. In extreme circumstances they will include tight planning restrictions and the diversion—deliberate or otherwise—of economic growth towards less favoured areas. The land use planning functions of government fall largely into this category, being used by authorities particularly during periods of economic growth to channel and to contain development pressures (Hall et al. 1973). During periods of economic recession, however, this regulatory function has been complemented by—or in some cases absorbed into—more active economic policies that look for ways of positively supporting enterprise, the workforce, and the wider community. Such *growth promotion* strategies involve a switch away from the management of overheating towards industrial promotion and economic development, with a commensurate switching of staff away from development control towards economic policy. Given enduring patterns of uneven development, with some regions growing quickly whilst others are restructuring or contracting, we can indeed find both approaches operating simultaneously across a country, with some authorities (typically in more prosperous areas) emphasising *growth management*, others (typically in less prosperous areas) engaging more directly and over the longer term in *growth promotion* (Collinge 1992, 1999; Collinge and Mawson 1994).

The uneven nature of economic growth across any territory means that it is possible, depending upon circumstances, to promote growth or to reduce contraction, in places through judicious promotional activities backed up with infrastructure initiatives, and through direct investment (whether foreign or domestic, whether private or public sector, and including capital spending by the authorities themselves). There are two aspects to growth promotion:

- In what might be called the *recession management* arm of this strategy, there are concerns to defend existing capacity in preparation for a future recovery in the economy. Capacity will be defended by supporting threatened sectors—manufacturing or services—through rent and tax relief, through assistance in addressing problems including financial problems or help with lobbying. Capacity will also be defended by supporting threatened workers with the preservation of

employment, with preservation and extension of skill levels (including the development of entrepreneurialism), and perhaps through intercessions with creditors of one sort or another (e.g. landlords) where this is possible. At the same time the social capital of the place must be preserved by supporting employment, encouraging employers—including the authority itself—to retain their workforce, and the maintain the general appeal of the area.
- In what might be called the *restructuring* arm of this strategy, the overall mix of the economy—its vulnerability and resilience—will be reviewed and efforts will be made to secure a sounder balance for the future. The sectoral or cluster focus will be upon certain nascent activities, upon building up social capital in the form of sectoral or cluster organisation, and (perhaps) upon intervention in the supply of entrepreneurs, venture, and development capital through (for example) advice and brokerage schemes of one sort or another. The other side of this coin will particularly focus upon skill development. It is possible to envisage the promotion of high value engineering skills and activities as contributing to this restructuring. Indeed, engineering skills and associated inward investment have been promoted over recent years as ways of reversing economic decline.

Needless to say, each of these strategies will be modified according to circumstances and pursued within the limits of authority powers and budgets.

5 Conclusions

Government at the national and other levels has responded to the requirements of business by providing or supplementing the social and economic conditions for economic development. It has also responded to the pressures associated with instability and unevenness by relating its economic policies to variations in performance over time and between locations. Strenuous efforts are made nationally, for instance, to moderate and to manage the business cycle in order to achieve sustained growth, and one way or another, authorities have been drawn into this process.

Efforts have also been made to support existing industries in depressed areas, to attract investment or customers to these areas, and to contain and channel growth in more prosperous parts of the country especially when there is a danger of overheating. In this chapter a framework has been presented to show the range of policy options that are available to governments and their agencies—whether liberal market or coordinated market economies—at different spatial scales, and during different episodes in their economic development. Depending upon the circumstances, these policies can be selected and targeted to achieve the promotion of high value engineering.

References

Collinge, C. (1992). The dynamics of local intervention: Economic development and the theory of local government. *Urban Studies, 29*(1), 57–75.

Collinge, C. (1999). Self-organisation of society by scale: A spatial reworking of regulation theory. *Environment and Planning D: Society and Space, 17*(5), 557–574. ISSN 0263-7758.

Collinge, C., & Mawson, J. (1994). The formation of local economic policies, Chapter 7. In B. Atkinson (Ed.), *Developments in Economics* (pp. 99–118). Ormskirk: Causeway Press.

Collinge, C., & Srbljanin, A. (2002). A network paradigm for Urban governance? In R. Hambleton, H. Savitch & M. Stewart (Eds.), *Globalism and Local Democracy: Challenge and Change in Europe and North America.* Basingstoke: Palgrave Macmillan. ISBN 0333772253.

Cooke, P., & Leydesdorff, L. (2006). Regional development in the knowledge-based economy: The construction of advantage. *The journal of Technology Transfer, 31*(1), 5–15.

Epstein, G. (2005). *Capital Flight and Capital Controls in Developing Countries.* Edward Elgar.

Etzkowitz, H., & Leydesdorff, L. (2000). The dynamics of innovation: From national systems and "mode 2" to a triple helix of university-industry-government relations. *Research Policy, 29*(2), 109–123.

Fukuyama, F. (1995). *Trust: The new foundations of global prosperity.* New York: The Free Press.

Hall, P., Thomas, R., Gracey, H., & Drewitt, R. (1973). *The containment of urban England* (Vols. 1 and 2). London: Political and Economic Planning.
Hall, P., & Soskice, D. (2001). *Varieties of capitalism: The institutional foundations of comparative advantage.* Oxford: Oxford University Press.
Harvey, D. (2007). *The limits to capital* (2nd ed.). London: Verso.
Jacobs, J. (1972). *The economy of cities.* Harmondsworth: Penguin Books.
Kenney, M., & Von Burg, U. (1999). Technology, entrepreneurship and path dependence: Industrial clustering in silicon valley and route 128. *Industrial and Corporate Change, 8*(1), 67–103.

Authors' Biography

Chris Collinge is Senior Lecturer in the Birmingham Business School. His research focuses upon regional economic change with particular reference to the emergence of knowledge-intensive manufacturing and services in a cross-national context. He has also researched evolving patterns of governance, strategy and leadership, and has written about scale as a tool for conceptualising spatial restructuring. He has published articles in Urban Studies, Environment and Planning A, C and D, Transactions of the Institute of British Geographers, Policy Studies and Local Economy journals. Recent books include Platforms of Innovation (2010), and Leadership and Place (2011).

Index

Y. Zhang and M. Gregory (eds.), *Value Creation through Engineering Excellence*,
https://doi.org/10.1007/978-3-319-56336-7

FSC
www.fsc.org
MIX
Papier | Fördert gute Waldnutzung
FSC® C083411

Kolibri 360 GmbH
Ferdinand-Jühlke-Straße 7
99095 Erfurt, Deutschland
produktsicherheit@kolibri360.de